Models and Modeling in Science Education

Volume 9

More information about this series at: http://www.springer.com/series/6931

John K. Gilbert • Rosária Justi

Modelling-based Teaching in Science Education

Springer

John K. Gilbert
The University of Reading
Berkshire, UK

Rosária Justi
Universidade Federal de Minas Gerais
Belo Horizonte, Brazil

ISSN 1871-2983 ISSN 2213-2260 (electronic)
Models and Modeling in Science Education
ISBN 978-3-319-29038-6 ISBN 978-3-319-29039-3 (eBook)
DOI 10.1007/978-3-319-29039-3

Library of Congress Control Number: 2016939958

Printed on acid-free paper

This Springer imprint is published by Springer Nature
The registered company is Springer International Publishing AG Switzerland

Foreword

From my collegiate experiences with and reading Gilbert's and Justi's respective research publications, I cannot imagine any two science education colleagues who are more suited to and qualified for writing a book entitled *Modelling-Based Teaching in Science Education.* Gilbert and Justi have a vast experience over more than two decades, collectively and independently, working with secondary science teachers in schools to implement a range of new teaching approaches and alternative curricula designed to improve students' learning outcomes. Their research with classroom teachers that includes the use of models, analogies, visualisation, and variations of assessment has been published in journals and in edited books. While all these research findings are accessible, it is a great advantage to science educators that ideas and findings from their research activities have been brought together within the extant literature under one cover.

This text is well structured and maintains a clear focus on the nature of models, modelling, and modelling-based teaching, thereby illustrating consistently that models are not only the basis of much scientific practice but also can – and should – play similar roles in teaching and learning school science. In this text, Gilbert and Justi provide considerable evidence that modelling play a central role in teaching and learning science but they also, rightfully, recognise the limitations of such teaching and explain what teachers can do to address these limitations.

This is a scholarly text and one that is eminently readable for university academics and also teachers. References are sourced from a wide informing literature not only from science education but also the history and philosophy of science and psychology. In this way, the authors situate their work in the past and current literature that is well synthesised such that there is a logical connectedness from the start to the end of each chapter and also from the start to the end of the book.

I have conducted classroom research with doctoral students and fellow colleagues on models, primarily used in chemistry teaching and analogies and metaphors used in science teaching, and examined the importance of different representations and modes of representations that incorporate visualisation in teaching and learning science. Consequently, many of these chapters are of personal interest to me. Notwithstanding my personal interests, what Gilbert and Justi have managed to do really well is to frame their own work in the extant literature; identify key issues that ensure success, or otherwise, of a particular teaching approach with the aspects of modelling and modelling-based teaching; and provide suggestions and recommendations for effective teaching and learning. The last point especially is why I believe that *Modelling-Based Teaching in Science Education* would also be a valuable resource for teachers interested in this style of enriched teaching with models. Furthermore, what additional research work is needed to enhance classroom practice of modelling-based teaching has also been presented.

Curtin University
Bentley, WA, Australia

David F. Treagust

Preface

The word 'model' in English is used in a wide variety of ways (OUP, 2008). A number of allied meanings are only found in everyday life:

- A garment made by a well-known designer. For example, a dress designed by Versace;
- A person who wears clothes to display them. For example, Kate Moss;
- A person who is a source of inspiration for a photographer or artist. For example, Joanna Hifferman and the painter Gustave Corbett;
- A person worthy of imitation. This is person who has achieved long-lasting heroic stature in a society. For example, Sir Edmund Hilary in New Zealand;
- An object worthy of imitation. This is an object that attracts emulators. For example, a vacuum cleaner designed by Sir James Dyson;
- An object that is smaller than the original. For example, the model of the Great Pyramid in Cairo Museum;
- A prototype of an object to be made in more durable material. For example, a clay model of a car made prior to its actual manufacture.

Other meanings are found both in everyday life and in science:

- A typical form or pattern. One example in each of the two contexts is: the basic layout of a passenger airliner; the array of glassware used in a chemical reaction;
- One object in a series of allied objects. One example in each of the two contexts is: a Mark 5 Volkswagen Gulf car, following on from Mark 1, 2, 3, 4; the electron cloud model of the atom, following on from the Thompson, the Rutherford, and the Bohr, models.

Yet other, overlapping, meanings have a particular status in science and technology:

- Objects that represent the original in a different scale aiming at supporting explanations and predictions about it. For example, a model of the HIV virus;

- A scientific description of something that is complex. For example, the Watson-Crick-Franklin model of DNA.

This wide range of meanings is very confusing to most people, particularly when they are learning or employing scientific ideas. In this book we are concerned with the wide range of scientific meanings contained in the latter two categories.

The great breadth and diversity of role of a model in science are captured in a typical (yet tautological!) definition of it as being a representation of things that are of interest to science. The formation and testing of models does play particular roles in science because they are concerned with the production of various types of explanation of the nature of the world-as-experienced. Thus ambition is far too demanding unless natural, complex, phenomena are simplified in some way. So this is done through the production and use of models. The particular importance of models and modelling in science is recognised, extensively if not always clearly, in the literature of the history and philosophy of science (for instance, in Hodson, 2009; Matthews, 2014).

Models can be placed into several types of category. Thus, although a model is always present in mental form in the mind of its inventor or subsequent user, it can take on one or more physical forms when placed in the public domain. These forms can be represented in a variety of media, for example, in the form of a gesture (e.g. of the relative position of objects), in a material form (e.g. a ball-and-stick representation of a crystal structure), in a visual form (e.g. as a diagram of a metabolic pathway), in a verbal form (e.g. an analogy for the structure of an atom based on that of the solar system), in a symbolic form (e.g. as a chemical equation), and in a virtual form (e.g. as a computer simulation). The range of entities that can be represented is wide: objects (e.g. of a virus), systems (e.g. of a blood circulation system), processes (e.g. of the liberation of energy from foodstuffs), events (e.g. of the attack of a white blood cell on a virus), ideas (e.g. of a vector of a force), and arrays of data about any of these entities.

For the purposes of this book, we define modelling as the dynamic process of producing, using, modifying, and abandoning the models in science. In the light of the wide range of meaning that the word 'model' has acquired, summarised above, it does seem that modelling is a core process in all human thinking and, as such, a vitally important focus for education.

In general, education has three broad aims. First, it is concerned with the transmission of socially valued knowledge across the generations such that the knowledge acquired by earlier generations is not lost. Second, it seeks to pass on the thinking skills that have produced that knowledge. Third, it supports the production of new knowledge through the use of these skills. The thinking skills involved in the conduct of science in particular are manifested in the processes that lead to scientific knowledge. Models and modelling, therefore, must play important roles in science education if the latter is to be 'authentic', that is to reflect how science has been and should be conducted (Gilbert, 2004).

The importance of models and modelling in the nature of thinking and in the history and philosophy of science has long been a matter of contention (for instance, by Giere, 1988). However, its saliency in discussions about science education has only gradually risen in the few decades or so. This process has several roots. The first was in the study of the meanings that students had for single words commonly used in science: the so-called misconceptions or alternative conceptions movement (Gilbert & Watts, 1983). This initially focused on the meanings held by students of individual words (for example, force, heat, light, energy). It gradually expanded to the study of how these meanings interacted, leading to understanding of complex phenomena by their integration into models, for example, of everyday movement, of the cooling of liquids, of the production of shades of colour, and of energy conservation (Gilbert & Boulter, 2000). The second root was the gradually emerging emphasis in curricula of the study of the nature and processes of scientific enquiry (Abd-El-Khalick, 2012). This perhaps occurred to some extent because of the need to provide a basis for the unification between the separate sciences – mainly physics, chemistry, biology, earth science – when these are amalgamated into 'general' or 'integrated' science courses in compulsory-age schooling. Models, being central to the history and philosophy of all the sciences, were seen as able to do this. The third role was the need to improve accessibility to the ideas of science, in the face of evidence that curricula had become overloaded with content, fragmented in structure, and too abstract, and divorced from phenomena of interest to students (Cerini, Murray, & Reiss, 2003). The outcome of these problems has lead to widespread student disengagement with the sciences. Particular models, applicable across diverse areas of content, were seen not only as potentially providing access to complex phenomena that are relevant to students' interests, as providing the basis for the integration of individual facts, and hence able to effect a simplification of the curriculum that made learning easier. The fourth root has been the advent of desktop computers with very large memory stores. These provide access to highly interactive 'modelling systems', thus enabling enquiry work focused on models and modelling to readily take place (Edelson, 2001).

This book has three purposes. First, it draws together, evaluates, and integrates the findings of the diverse literatures that have contributed to current knowledge of the overall field of modelling in science education. Second, it justifies the central contribution of modelling to science curricula. Third, it identifies the research and development work still needed for that contribution to be realised in classroom practice.

As such, the book has six overlapping audiences:

- Curriculum designers, for it is they who have the best opportunity to signal the importance of modelling to teachers;
- Public examiners, for it is they who define what knowledge of modelling can be validly and reliably assessed;
- Textbook designers, for it is they who translate the intentions of curriculum designers and public examiners into forms readily grasped by students (and their teachers!);
- Teacher educators, for it is they who have the best opportunity to introduce pre- and in-service teachers to the potentialities and realities of modelling;

- Advanced students of science education and curriculum design, for they have the opportunity to study the ontological and epistemological bases of modelling;
- Science education researchers, for they have the task of filling the gaps in our understanding of modelling.

The book has 12 Chapters. They are successively concerned with:

1. The challenges that science education currently faces, together with the assertion that an education in and about modelling can help meet these challenges.
2. The notion of 'model' and the knowledge and skills that contribute to the production and validation of models.
3. The notion of 'authentic learning in science' together with an evaluation of how modelling can contribute to that authenticity.
4. An exploration of the meaning of 'modelling-based teaching' together with the presentation of an approach based on the 'Model of Modelling'.
5. As the meanings of the words 'concept' and 'model' are often confused in the literature, an exploration of the scope and limitations of both is conducted.
6. The use of argumentation in the acts of creating and validating models.
7. The contribution that 'visualisation' makes to the creation of models.
8. The central role of analogies in modelling-based teaching.
9. The way that modelling contributes to the core curricular aims of 'understanding the scientific enterprise'.
10. The structure of a learning progression for modelling.
11. The professional development of teachers needed to implement modelling-based teaching.
12. A review of the lacks of definitive knowledge needed for the universal implementation of modelling-based teaching together with suggestions about how this situation might be addressed.

We decided to write this book because, although we have jointly and singly written about models and modelling for over 20 years, we felt to need construct an overarching view of the field. At the same time, the place of 'modelling' in national mandatory curricula was being progressively strengthened and we felt that science educators in general would value such an overview. Although based in different continents and having very different professional commitments, we did manage to meet at least twice in each of the 3 years that it took us to write. Nothing could have been achieved without e-mail, Skype, and the generous support of our professional friends, especially Ana Sofia Afonso, David Treagust, Izabella Martins, Matthew Newberry, Maurice Cheng, Nilmara Mozzer, Paula Mendonça, and Poliana Maia. We view the outcome as 'work in progress', for Chap. 12 sums up the serious gaps

in knowledge that currently exist. Having toiled through many hundreds of papers, we would respectfully suggest that future authors define their terms and write with an eye to classroom implications.

Mayford, Woking, UK John K. Gilbert
Belo Horizonte, Minas Gerais, Brazil Rosária Justi

References

Abd-El-Khalick, F. (2012). Nature of science in science education: Towards a coherent framework for synergistic research and development. In B. J. Fraser, K. G. Tobin, & C. J. McRobbie (Eds.), *Second international handbook of science education* (pp. 1041–1060). Dordrecht, The Netherlands: Springer.

Cerini, B., Murray, I., & Reiss, M. (2003). *Student review of the science curriculum: The consultation process*. London, UK: Planet Science.

Edelson, D. C. (2001). Learning for use: A framework for the design of technology-supported enquiry activities. *Journal of Research in Science Teaching, 38*(3), 355–385.

Giere, R. N. (1988). *Explaining science: A cognitive approach*. Chicago, IL/London, UK: University of Chicago Press.

Gilbert, J. K. (2004). Models and modelling: Routes to a more authentic science education. *International Journal of Science and Mathematics Education, 2*, 115–130.

Gilbert, J. K., & Boulter, C. J. (Eds.). (2000). *Developing models in science education*. Dordrecht, The Netherlands: Kluwer.

Gilbert, J. K., & Watts, D. M. (1983). Conceptions, misconceptions, and alternative conceptions: Changing perspectives in science education. *Studies in Science Education, 10*(1), 61–98.

Hodson, D. (2009). *Teaching and learning about science: Language, theories, methods, history, traditions and values*. Rotterdam, The Netherlands: Sense.

Matthews, M. (Ed.). (2014). *International handbook of research in history, philosophy and science teaching*. Dordrecht, The Netherlands: Springer.

OUP. (2008). *Concise oxford english dictionary*. Oxford, UK: Oxford University Press.

Contents

Chapter 1
Facing the Challenges to Science Education in Schools: The Contribution of Modelling

Abstract The grounds are laid for the advocacy of an increased role for modelling in science education. Anecdotal evidence of students' lack of engagement in science classes is used to support widespread dissatisfaction by governments with students' levels of attainment in international assessments and with their disinclination to continue to study the discipline after the years of compulsory schooling. The underlying causes are attributed to: the heavy content load, often presented within a curriculum that is antiquated and rigidly structured; to problems over the supply of suitably qualified teachers; and to the, often excessive, adoption of didactic methods of teaching. Efforts to attain 'scientific literacy for all' are seen as likely to overcome these problems. The achievement of the ability by students to engage in modelling is seen as a major contributor to the attainment of this goal.

The national educational authorities throughout the world currently perceive school science to be facing a number of challenges. Some of these are of long standing. Others are of a more recent origin, stemming from changes in policy about what should be learnt, how, and by whom. Both sets of challenges will have to be addressed if the quality and extent of learning of science and about science are to be steadily improved. In this chapter, we discuss the nature and origin of these challenges, which overlap in their natures and causes. We then go on to discuss the ways in which they can be addressed, and outline the contributions that modelling can make to that process.

The Nature of Long-Standing Challenges

There seems to be a general belief that science education at school level is not as successful as it ought to be. However, whether patterns of attainment are perceived to be acceptable or not does, of course, depend on how they are measured and what comparisons are made.

J.K. Gilbert, R. Justi, *Modelling-based Teaching in Science Education*, Models and Modeling in Science Education 9, DOI 10.1007/978-3-319-29039-3_1

Many countries now participate in the assessment of students, most notably those that produce the comparative 'Trends in International Mathematics and Science Study' (TIMSS) reports. These are used by the educational authorities in individual countries to evaluate the relative standards of attainment of science education in their government-funded schools when compared with those in other countries. The education authorities of South Korea, Singapore, and Hong Kong must be gratified to see that their systems consistently fill the top few places for science in respect of 10 year olds and for 14 year olds. Other countries, perhaps inevitably, will be less satisfied. For example, in the case of the 10 year-old cohort, England dropped from 7th to 15th position between 2007 and 2011, whilst the 14 year-old cohort fell from 5th to 9th position (BBC News, 2013; Martin, Mullis, Foy, & Stanco, 2012). But does any such 'slippage' suggest that science teaching is less successful than in years gone by? Perhaps not, for – to continue with the example of England – the National Testing Programme showed that there was a 19 % improvement in respect of prescribed levels of attainment by 10 year-olds, to about 90 %, between 1999 and 2009, the comparable figures for 14 year-olds being 16 % to about 75 % (The Royal Society, 2010). The conclusion that may be drawn is that, whilst some countries are progressing faster than others in respect of the learning of that knowledge assessed by the TIMSS tests, and there is always some scope for improvement against intra-national standards, general progress may well be taking place. The issue is then what evidence is there to support the belief by the general public, or perhaps rather by the media seeking 'negative' stories, that things are not going as well as they might?

One source of concern is the perception, reported anecdotally by many teachers in the UK at least, that some of their students, for some of the time at least, are not mentally engaged with the themes of lessons. Indeed, the word 'some' in the sentence above can range in meaning from 'a few' to 'almost all'. Consistent with the findings of studies in many countries and over many years, Mortimer and Scott (2003) showed that science teachers in the UK dominate the talk in science classrooms, with students having relatively little opportunity (and hence little mental initiative) to actively and creatively contribute to the proceedings. This lack of engagement may lead to, or stem from, poor general attitudes of students towards science education. With the possible partial exception of people from the richest countries in the world and from females, there are positive attitudes towards science and technology as such among young people (Sjøberg & Schreiner, 2010). However, attitudes to science education are much more negative (Sjøberg & Schreiner, 2010), the subject (or subjects) of science being seen as less interesting than other subjects, as not arousing curiosity, and not always leading to improved career opportunities.

One consequence of negative attitudes, where they exist, is that students do not opt to continue their post-compulsory studies of science (unless, perhaps, in periods of economic depression) in as great a number as governments seem to wish. Why is this? The Aspires Project in the UK has recently identified three factors in reasons given by 10–13 year-olds (the age at which such decisions begin to be taken) for not pursuing a career that requires the further study of science. These are that: many of their families do not have much 'science capital' (science qualifications or social contacts with scientists or engineers) and hence are unable to see the potential of such a career; science-related careers are seen as limited in number and only suitable for the most 'brainy' among them; science careers are seen as 'masculine' in expectation (Archer, Osborne, & Fortus, 2012). The last reason focuses attention on the relatively low voluntary engagement by girls in many – but not all – countries. For example, only 49 % of state co-educational schools (the dominant social arrangement) in England had girls studying physics at post-compulsory level in 2012 (Institute of Physics, 2012). This particular outcome seems to be the consequence of peer pressure ('physics is not feminine'), the perception that scientists are bearded men in laboratory coats, parental views on the employment potential of physics, and the reputation of physics as being hard to learn.

This inter-connected series of problems and challenges stems to some extent from what is taught in science and how it is taught. There seems little doubt that the school science curriculum in most countries contains too much content: phenomena, individual facts, theories, and concepts. New content is constantly added, usually belatedly reflecting major advances in science *per se*, but little if any is removed. The load is too great, to fragmented for meaningful learning by many students, and not of the type that ought to lead to the application of ideas in novel situations. Moreover, the material is too often presented to students within the exemplar circumstances in which it was originally explored. These circumstances are too often far removed from those likely to be encountered by today's students. Adey (1997) has argued that learning has taken place within contexts that are familiar to students is more likely to be effective, retained, and subsequently used. The criteria for identifying contexts that are suitable for a given learning task in science have been explored and their value established in promoting learning than can be transferred to other contexts (Gilbert, Bulte, & Pilot, 2011).

Even where contexts for learning science can be identified that ought to be conducive to more widespread and effective learning, the structure of the curriculum, allied to the educational background of the teachers, may conspire to defeat success. The science curriculum for compulsory-age students is organised in most countries either as separate sciences (typically 'physics', 'chemistry', 'biology') or in some amalgam of those (typically 'modular', 'general' or 'integrated' science). The problem is that most science teachers, even those working in secondary (high) schools, only get an in-depth education in one, maybe two, science subjects. They are too often required to teach outside their 'comfort zone', leading to a more traditional, didactic, approach to teaching. The overall consequence of this mismatch too often seems to be that any aspirations for a more radical approach to science education is progressively 'watered-down' as they are implemented in the classroom: in short, little changes (van der Akker, 1998).

The most radical approach to meeting all these challenges may be to pay more attention to the 'student voice' – what young people say they want to learn and how they want to learn it (Jenkins, 2006). It is, after all, the rapidly changing world of tomorrow in which they will have to conduct the majority of their lives. However, older generations will be reluctant to yield power over the nature of the science curriculum – for good or ill – and so we must look to see how these challenges can be met within existing structures of management.

Addressing Long-Standing Challenges

Attempts to address the challenges outlined above have taken place severally and in many countries over many years. A consensus on what to do gradually emerged, for example as codified in the policy document sponsored by UNESCO (Fensham, 2008). Although they are interlocking, the measures that can be taken are capable of presentation separately. These are in respect of: the approaches to teaching and learning adopted in the classroom; the background and training of teachers; the purpose and structure of the curriculum. They will have an impact cumulatively on both the level of student attainment, on attitudes to science education, and hence on career aspirations. Taking these separately:

The Approaches to Learning and Teaching Adopted in the Classroom

The keys to improving student attitudes to science education, as manifest in their engagement in classes and their consequent attainment, lie in the adoption of suitable approaches to learning and teaching. For many years the dominant assumptions behind all teaching and learning was that, until they were 'instructed', students knew nothing about most topics on the curriculum. The teacher's task was to present valued knowledge accurately such that all the students acquired an accurate 'copy' of that knowledge. The continued existence of the challenges, outlined above, is a testimony to the invalidity of those assumptions. In the last few decades, this *transmission* model of teaching and learning has been superseded, at least in the ideology of science education, if not completely in its practical implication, by several *constructivist* models for pedagogy. These assume that what is learnt is related to what is already known about the topic, this process giving the learner a sense of ownership of the knowledge acquired, even if this knowledge is often partially or greatly inaccurate (Gilbert & Watts, 1983). The model being most actively advocated at the moment is that of *social constructivism,* in which what is learnt is heavily influenced by social interactions with the teacher and other students (see Gilbert, 2013 for a summary).

The overall implications of the social constructivist model of learning and teaching in science education have been researched into over many years (see Scott, Asoko, & Leach, 2007 for a review). It is not appropriate to synthesise that research literature extensively here, but instead to point to some key consequences of its use that are important in the discussion of modelling that is the focus of this book.

It does seem that, if the adoption of a social constructivist approach to teaching and learning is extensive across the different subjects, the general 'school climate' in such an institution is one that does lead to improved attitudes and attainment in science (Olitsky, 2007; Vedder-Weiss & Fortus, 2012). One of the major components of a positive school climate is that it enables students to articulate and augment their 'cultural capital' that is relevant to the study of science, for students from different socio-economic and ethnic backgrounds have access to:

> unequal knowledge about courses and the careers they lead to, the cultural models which associate certain occupations and certain educational options and certain cultural option with a particular social backgrounds, and the socially conditioned predisposition to adapt oneself to models, rules, and values which govern the school system (Bourdieu & Passeron, 1979, p. 13).

This development of cultural capital, manifest in a rising (rather than declining) interest in science, is associated with the increased levels of personal mental activity with the subject(s) (Swarat, Ortony, & Revelle, 2012) (see Fig. 1.2). This mental activity, arising from or leading to, an associated emotional engagement (Lin, Hong, & Juang, 2012) augments students' sense of 'personal identity' (Archer et al., 2010). However, these gains in emotional engagement, physical and mental activity, self-identity, and hence achievement, depend critically on the leadership shown by the teacher (Odom, Stoddard, & LaNasa, 2007). Thus the background and training of science teachers are of crucial importance. The consequences of as successful address to the challenges faced are summarised in Fig. 1.1.

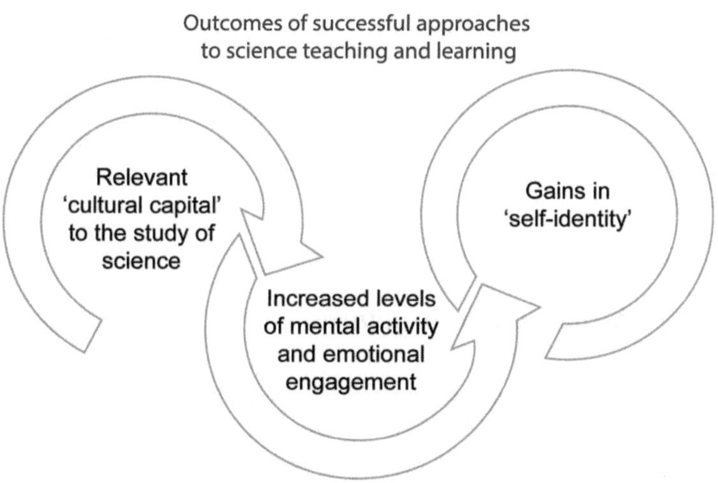

Fig. 1.1 Aspirations for the provision of science education

The Background and Training of Teachers

Primary (elementary) teachers in most countries are generalists, having responsibility for the education of their pupils in respect of all the school subjects. This means that they are unlikely to have themselves specialised in science at school, university, or in teacher training. Secondary (high school) teachers will almost certainly have studied one or more of the sciences to some extent at university, but the content matter of these courses may not coincide with what is expected of them as science teachers. These circumstances point to a need for the professional education of all teachers of science not only to be as an efficient address to the challenges outlined above but also to continue throughout their professional lives.

This education will have to be focused on three major themes. First, the recognition that their subject knowledge may be inadequate for the teaching tasks that they face. Indeed, they will probably share some of the 'alternative conceptions' shown by their students (Gilbert & Watts, 1983). Second, that their 'pedagogic content knowledge' (Shulman, 1987) – the complex base of the strategies that they use to support the acquisition of knowledge by their students – will need sustained development. For example, how to identify and address suitable contexts in which science and technology are manifest, and how to produce and conduct activities that may result in students' learning of the relationships between science and technology. Third, that their personal beliefs and attitudes towards teaching and learning, usually based on own educational experience, will need to be identified and revisited in order to ensure that they are commensurate with the successful implementation of the reforms that are currently being adopted (Gilbert, 2010). In Chapter 11 we show how these might be addressed, over some years, in respect of the topic of 'modelling'.

The introduction of new approaches to teaching and learning, made possible by changes in teacher education, are manifest within the framework provided by the overt purposes and structure of the school science curriculum. It is to changes in respect of these that we now turn.

Curriculum Purposes and Structures

Science was first introduced into the school curriculum, in Western Europe and North America, in the late 1800s. The purpose was to identify those students who seemed suited to the study of the broad field and to provide them with an understanding of those core, usually abstract, concepts on which they could build at university level. This broad purpose was maintained even though increasing numbers of students had no aspiration to study science or technology at university level, or even to have a university degree. The students who have not been interested in science – who rapidly became the majority – might be interested in the applications of science to everyday life. This motivated the emergence of a new set of purposes. Alas, it is only in recent years that the tensions between these two sets of purposes have become very evident in the structures adopted for the science curriculum. In attempts to accommodate these two models of the purposes, a wide variety of curriculum strategies have been adopted. The content matter may still be structured in terms of single subjects labelled 'physics', 'chemistry', biology'. The subjects may be studied sequentially, for example one in each of successive school years. Alternately, they may be broken up into distinct 'modules'. Here chunks of content drawn from all three subjects are dealt with separately and also usually sequentially. The totality of the content may be the same for all students and be interspersed in such an 'integrated science' approach. All students may be required to progress through the same content at the same pace, or the curriculum provided may differentiate into 'streams' as students move up the school.

Our central argument in this book is that all this range of different approach can be accommodated within an approach to curriculum design that makes appropriate use of the concept of 'modelling'. Most importantly, such an approach can accommodate the notion of 'scientific literacy', which avoids the making of premature decisions by students about their interests, ambitions, and possible future careers. It is to the notion of 'scientific literacy' that we now turn.

Facing the Challenge of Tomorrow: Scientific Literacy for All

There has been a general consensus, for over a decade within national educational authorities, for instance those of Canada (The Council of Ministers of Education, 1997), Australia (Australian Curriculum Assessment and Reporting Authority, 2010), USA (National Research National Research Council, 2012), that the purpose of science education is to support the attainment of 'scientific literacy' by all compulsory school age students. The difficulties in deciding both what this goal entails and in evaluating progress towards it have rested on an inability to agree on what 'scientific literacy' actually means. Many analytical reviews of the field have been produced (for example, by Laugksch, 2000; Roberts, 2007).

Arriving at a usable interpretation of the phrase 'scientific literacy' involves splitting it into its component parts. In brief, the notion of 'scientific' involves amalgamating ideas about the nature of science, the nature of technology, the relationship between the two, whilst that of 'literacy' involves the capacity to interpret text, whether written, spoken, or otherwise represented. Having looked at the two components separately, we can then discuss what should be involved in an education towards scientific literacy and, finally, the place of modelling in that.

The 'Scientific' Aspects of Scientific Literacy

The social enterprise that is science seeks to produce explanations of the world-as-experienced that are closely related to the conduct of empirical enquiry and to the production of evidence from that enquiry. Much has been written about what this assumes – 'nature of science' – over the last few 100 years. It is not our intention to recapitulate that intriguing journey. Rather, we will use an authoritative secondary source (Rutherford & Ahlgren, 1990) to set out some of the core assumptions that science makes. These are that the activity is based on:

- *Realism*. The world is real, that is, it exists independently of human experience of it. This implies that reliable and persistent knowledge of the world-as-experienced can be obtained;
- *Objectivity*. Science explicitly attempts to be objective, that is, to avoid personal bias in what is seen and done. These attempts are intended to ensure that reliable knowledge can be built up by involving a number of different scientists and the resources at their disposal;
- *Social shaping*. The identities of the phenomena that the science community chooses to enquiry into are influenced by cultural factors. Such a choice is made, to some extent, on the basis of what aspects of the world-as-experienced seem particularly salient within a given cultural tradition at a particular time. This means that perceptions of social issues as being fundamentally important (for example 'climate change') or as the results of the activities of pressure groups (for example 'particle physicists') play an important role in establishing priorities;
- *Physical entities*. Science can only explain phenomena that depend of the presence of physical entities. This means that a large range of social phenomena based primarily on belief rather than on physical and reproducible evidence cannot be investigated scientifically. These beliefs lie along a spectrum, ranging from simple folk-wisdom superstitions to those represented in codified religions. However, over the years, many phenomena once thought of as superstitions have been successfully investigated scientifically (for example, water divining, acupuncture);
- *Methodological pluralism*. There is no one single methodology of enquiry used by science, rather a 'tool box' from which an approach is drawn. For example, enquiries into genetic phenomena rest heavily on direct intervention into them, whilst enquiries into astronomical phenomena cannot do so and rest on the chance arrival on Earth of evidence of the occurrence of celestial events in the distant past. Very different empirical procedures are involved in such diverse enquiries, but the mental frameworks within which they rest are of a common type. In our view, this substantially, but not completely, refutes the notion that there is one, universal, way of 'being scientific'. Talking about *the* methodology of science is thus perhaps too simplistic, if semantically convenient;
- *Logicality*. The conduct of science depends heavily on the use of logic. A reliance on 'if–then' arguments presumes that all events have causes. Thus, science presumes that both the operation of particular causes and their consequences for associated events can be precisely observed, either deterministically or probabilistically;
- *Predictability*. The production of predictions, the projection of the behaviour of a phenomenon under different circumstances from those in which it is initially encountered, is a key aspect of science. The key issue here is the *precise* nature

of the expectations that can be arrived at: the magnitudes of actual measurements to be obtained must be accurately anticipated;

- *Tentativeness*. The explanations that science produces are always open to critical evaluation. This can lead to a cycle of their acceptance, modification, and replacement, or to their abandonment. If the assumptions on which a set of predictions are based are faulty, then the predictions themselves will not be borne out by empirical observation. A slight discrepancy between the two can lead to the assumptions being adapted before the cycle is repeated. However, a wide discrepancy can lead to a decision to 'start again from scratch'. Scientists sometimes speak as if their current theories fully represented phenomena. In our view, this is only for linguistic convenience, for they should know, from experience, that it is unlikely to be justified;
- *Persistence*. Many of the ideas that science produces remain in use for many years (for example Newton's Laws of Motion). This is mainly because they produce explanations of certain phenomena that are acceptable for the uses to which they are commonly put, and because they support predictions that guide future studies.

Technology, on the other hand, seeks to meet everyday needs and for wishes to be realised. It does so by the activity of engineering, of manufacturing objects, creating systems, and subsequently using the objects in those systems. Pacey (2007) modelled the practice and outcomes of engineering and technology as consisting of three components: the technical aspect, that is, the knowledge, skills, resources, that are employed; the organisational aspect, that is, the social arrangements for production and distribution of its practice and outcomes; and the cultural aspect, that is, the goals and values underlying decisions on what technologies to produce and use. Both the practice and outcomes of engineering and technology are closely related to those of science, whether the two activities take place either sequentially or concurrently (Gardner, 1994). Whilst technologies may be initially produced, they become more effective when the underlying scientific explanation for the phenomena on which they are based become known, for example the creation of specialist steels after the causes of the strength of iron were scientifically understood. If specific scientific explanations are known, their use in the creation of technologies may follow later, for example the properties of silicon were known long before microprocessors were made. In some cases, scientific enquiry is conducted in order to lead to preconceived technologies, for example establishing the detailed behaviour of genes in order to perform gene therapy.

The 'Literacy' Component of Scientific Literacy

Being scientifically literate means not only understanding the processes and outcomes of science and engineering, but also understanding the complex knowledge that underlies the ways in which these are communicated, for example through the use of writing, tables, graphs, diagrams. These complexities have their causes in the nature of language itself and how it is used in science. Osborne (2002) has identified four key types of complexity in the language used by science that make the attainment of scientific literacy challenging. The first of these stems from the fact that science is polysemous: words used in everyday language have special meanings when used in science, for example 'force', 'reaction', 'hybrid'. The second is that logical connectives have a great impact on the meaning of statements in science e.g. the inclusion of 'and', 'or', 'either–or', 'implies', 'if–then', 'however'. The third is that the language of science is multi-semiotic: neologisms are created, for example 'electron', 'allotrope', 'gene', whilst diagrams, tables, and graphs make their own distinctive contributions to the communication of meaning. The fourth is that particular genres of representation are adopted by science when communicating its procedures or outcomes. Most well known of these genres is the use of the 'passive voice', in which the actors, as well as the contexts of and the purposes for actions, are removed, leaving a uniquely sterile form of narrative.

Acquiring the 'literacy' component of 'scientific literacy' is clearly a complex and demanding task. Kintgen (1988) proposed a four-level model for the development of literacy in general that can be adapted into the context of scientific literacy. The 'signature' level, only being able to read and write one's own name, results in a challenge to those who have the lowest attainment in science education. The second level is that of 'recitation', where the words used in science can be read or spoken, but without any understanding of their meaning. The demonstration of such a level will prove very frustrating for a science teacher, although a student who shows it can often score well on the simpler types of multiple choice examination questions. The third level is that of 'comprehension', in which a person can use a sound knowledge of concepts to understand an unfamiliar scientific text. This is surely what must be aimed at for all as a minimum competence. A fourth, or analytical, level would enable the reader to go further and draw inferences from what is read. This would allow a person to apply scientific knowledge to understand distinct phenomena and/or contexts.

These levels of scientific literacy will be manifest both in specific contexts and for particular purposes. Shen (1975) identified three such distinct contexts and their allied purposes. On this scheme, the first type is the 'practical' context. This is the tackling of everyday problems, for example the maintenance of health, the use of domestic technologies. The contexts and purposes of the second type of scientific literacy are termed the 'civic'. The 'civic' context and purposes deal with the taking of decisions by societies, locally (where to locate a garbage incinerator), or regionally (the route of a new road), or nationally (whether to adopt nuclear power), or globally (what to do about global warming). The 'cultural' type of scientific literacy refers to fundamental issues in humanity's perception of reality, such as the structure of the universe, the origin and evolution of life on Earth. The mere possibility that a cultural type of scientific literacy exists demonstrates the great achievements of science in recent centuries, these having being made possible by the provision of sophisticated instrumental technologies (like the radio telescope, and the electron microscope), and the application of much thought and enquiry.

The demands of the 'practical' type must provide the basic contexts and purposes in which particular levels of attainment of scientific literacy are aspired to. The nature and complexity of the education provided for scientific literacy of the 'practical' type will expand and deepen with progress through the school. In doing so, students' knowledge and skills will develop. This will go hand-in-hand with the progressive address to 'civic' contexts and 'cultural' contexts. However, if it serves the purpose of maintaining and extending students' interest and capabilities in science and technology, these three categories of context may, of course, be addressed in parallel.

Our central argument in this book is that 'modelling' can play a central role in the provision of an education for 'scientific literacy'.

The Role of Modelling in an Education for Scientific Literacy

Derek Hodson (2009) has identified the component topics, concepts, and skills that would form the basis of a universal curriculum aimed at supporting the development of scientific literacy. Meeting the needs of possible future specialists in science, engineering, and technology, involves an extension and elaboration of that curriculum in terms of those components. The core argument in this book is that learning about modelling and developing the skills to engage in modelling must play major roles in such provisions. The way that it can do so falls under four distinct headings:

Modelling Can Provide a Way to Reconstruct Established Scientific Models

A focus on established scientific models provides a route to the reduction in the content of both the curriculum for all students. Each model of the limited number of accepted models that must be addressed provides, by its very nature, an overarching representation of a variety of separate facts manifest in a wide variety of contexts. This reduction in content will allow students to engage more readily in meaningful learning, for they will have the mental capacity to engage in problem solving using these models. Assuming that the models reconstructed provide explanations of situations to which students can readily relate, their engagement with the process and content of their science education could improve. This acquisition of the skills of modelling would provide them with a sense of ownership of their knowledge, for they can use such skills to inquire into many of the phenomena that they meet in their everyday lives.

Modelling Will Be Recognised as a Core Component in the Conduct and Validation of Science and Technology

In the process of acquiring the knowledge and skills involved, students would come to appreciate the central roles played by models and modelling in the creation and validation of scientific and technological knowledge. When these capabilities are developed, students will be able not only to interpret data that are claimed to be scientific but also to evaluate any claims that are made about the significance or use of that data. These capabilities are the core components of scientific literacy and are particularly important where scientific models are used as the basis for the design of technological artefacts. The integration of the knowledge and skills involved will enable the relationship between science, engineering, technology, and their applications and implications not only for the environment but also for society, to be appreciated in depth.

Modelling Can Be a Route to the Development of General Mental Skills

Foremost of the skills involved in modelling is an understanding of the language that is used in the conduct of science, engineering and technology: the acquisition of the 'literacy' component of 'scientific literacy'. This is so important because language is the medium by which ideas are created and passed on to others. Hence an engagement in modelling will lead to an enhanced ability to think and to communicate in general and not only in respect of phenomena to which the methodology of science readily relates. This clear expression of scientific, engineering, and technological ideas is a vital component in establishing the worth of such knowledge.

Modelling Entails a Further Development of Personal Values Concerning the World-as-Experienced

As has been already said, the acquisition of the skills of modelling can provide the central tools with which students can conduct scientific enquiries about the world-as-experienced. In doing so, they are equipped to enter into scientific debates: their attitudes to science education should become more positive. Scientific knowledge should then be seen as the outcome of human argument and agreement, not just as a 'rhetoric of conclusions' the bases for which are not understood. They will see scientific debates as activities which they can, at best, meaningfully engage in, and, at worst, understand. The use of these skills should also sensitise students to the nature and importance of the socio-cultural circumstances in which scientific enquiry has, does, and will take place. Such a sensitisation will inevitably lead to a greater awareness of the ethical issues surrounding the conduct of scientific, engineering and technological work and to the values associated with their applications and implications.

Conditions for Success

If the major contribution of modelling to education for scientific literacy for all and for the education of future specialists is to be realised, a number of conditions must be met. The mental activities, the physical and social circumstances, conducive to the development of these skills must all be fully appreciated by curriculum designers and teachers and acted upon. This will involve a substantial shift in current patterns of classroom interaction and indeed in the general organisation of science education. Meeting these conditions will depend on science teachers having the appropriate beliefs, attitudes, and knowledge. This book is an attempt to unpick the conditions for success in respect of all these conditions. Their interlocking nature is represented in Fig. 1.2.

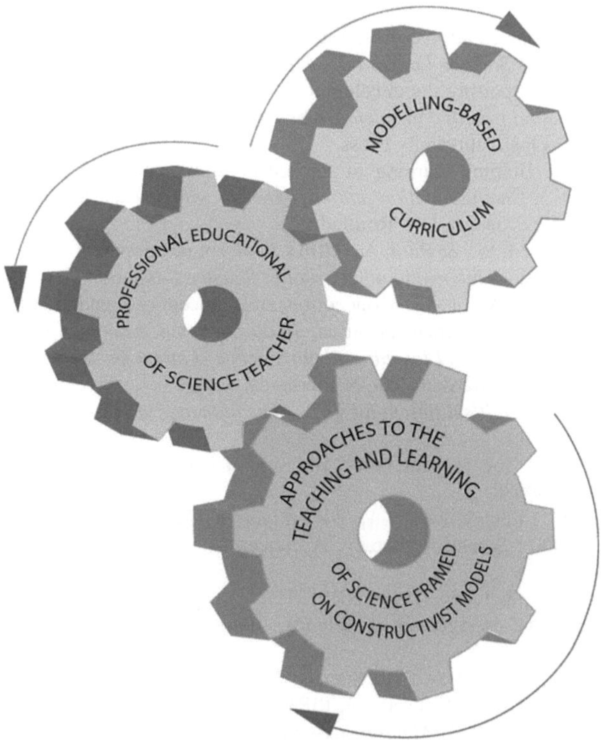

Fig. 1.2 Avenues to more successful science education

References

Adey, P. (1997). It all depends on the context, doesn't iI? searching for general educable dragons. *Studies in Science Education, 29*(1), 45–92.

Archer, L., DeWitt, J., Osborne, J., Dillon, J., Willis, B., & Wong, B. (2010). "Doing" science versus "being" a scientist: 10/11-year-old schoolchildren's constructions of science through the lens of identity. *Science Education, 94*(4), 617–639.

Archer, L., Osborne, J., & Fortus, D. (2012). *Ten science facts and fiction: The case of early education about STEM careers*. London, UK: The Science Council.

Australian Curriculum Assessment and Reporting Authority. (2010). *Science: Foundation to year 10 curriculum*. Canberra, Australia: Australian Curriculum, Assessment and Reporting Authority.

BBC News. (2013). Asian top school tables. *BBC News: Education and Family*. http://www.bbc.co.uk/news/education-20664752%5D

Bourdieu, P., & Passeron, J. (1979). *The inheritors*. Chicago, IL: The University of Chicago Press.

Fensham, P. (2008). *Science education police-making: Eleven emerging issues*. Paris, France: UNESCO.

Gardner, P. (1994). Representations of the relationships between science and technology in the curriculum. *Studies in Science Education, 24*(1), 1–28.

Gilbert, J. K. (2010). Supporting the development of effective science teachers. In J. Osborne & J. Dillon (Eds.), *Good practice in science teaching: What research has to say* (pp. 274–300). Maidenhead, UK: Open University Press.

Gilbert, J. K. (2013). Helping learning in science communication. In J. K. Gilbert & S. M. Stocklmayer (Eds.), *Communication and engagement with science and technology* (pp. 165–179). New York, NY/London, UK: Routledge.

Gilbert, J. K., Bulte, A. M. W., & Pilot, A. (2011). Concept development and transfer in context-based science education. *International Journal of Science Education, 33*(6), 817–837.

Gilbert, J. K., & Watts, D. M. (1983). Conceptions, misconceptions, and alternative conceptions: Changing perspectives in science education. *Studies in Science Education, 10*(1), 61–98.

Hodson, D. (2009). *Teaching and learning about science: Language, theories, methods, history, traditions and values*. Rotterdam, The Netherlands: Sense.

Institute of Physics. (2012). *It's different for girls – The influence of schools*. London, UK: Institute of Physics.

Jenkins, E. W. (2006). The student voice and school science education. *Studies in Science Education, 42*(1), 49–88.

Kintgen, E. R. (1988). Literacy literacy. *Visible Language, 1*(2/3), 149–168.

Laugksch, R. C. (2000). Scientific literacy: A conceptual overview. *Science Education, 84*(1), 71–94.

Lin, H.-S., Hong, Z.-R., & Juang, T.-C. (2012). The role of emotional factors in building public scientific literacy and engagement in science. *International Journal of Science Education, 34*(1), 25–42.

Martin, M. O., Mullis, I. V. S., Foy, P., & Stanco, G. M. (2012). *TIMSS 2011 international results in science*. Chestnut Hill, MA: TIMSS & PIRLS International Study Center.

Mortimer, E. F., & Scott, P. (2003). *Meaning making in secondary science classrooms*. Maidenhead, UK: Open University Press.

National Research Council. (2012). *A framework for K-12 science education: Practices, crosscutting concepts, and core ideas*. Washington, DC: The National Academies Press.

Odom, A., Stoddard, E., & LaNasa, S. (2007). Teacher practices and middle-school science achievements. *International Journal of Science Education, 29*(11), 1329–1346.

Olitsky, S. (2007). Promoting student engagement in science: Interactions rituals and the pursuit of a community of practice. *Journal of Research in Science Teaching, 44*(1), 33–56.

Osborne, J. (2002). Science without literacy: A ship without a sail? *Cambridge Journal of Education, 32*(2), 203–218.

Pacey, A. (2007). *The culture of technology*. Oxford, UK: Basil Blackwell.
Roberts, D. (2007). Scientific literacy/science literacy. In S. K. Abell & N. G. Lederman (Eds.), *Handbook of research on science education* (pp. 729–780). Mahwah, NJ: Lawrence Erlbaum.
Rutherford, F., & Ahlgren, A. (1990). *Science for all Americans*. Mahwah, NJ: Lawrence Erlbaum.
Scott, P., Asoko, H., & Leach, J. (2007). Students conceptions and conceptual learning in science. In S. K. Abell & N. G. Lederman (Eds.), *Handbook of research in science education* (pp. 31–56). Mahwah, NJ: Lawrence Erlbaum.
Shen, B. S. P. (1975). Science literacy. *American Scientist, 63*(3), 265–268.
Shulman, L. S. (1987). Knowledge and teaching: Foundations of the new reform. *Harvard Educational Review, 51*(1), 1–22.
Sjøberg, S., & Schreiner, C. (2010). *The ROSE project: An overview and key findings*. Oslo, Norway: University of Oslo.
Swarat, S., Ortony, A., & Revelle, W. (2012). Activity matters: Understanding student interest in school science. *Journal of Research in Science Teaching, 49*(4), 515–537.
The Council of Ministers of Education. (1997). *Common framework of science learning outcomes K to 12: Pan-Canadian protocol for collaboration on school curriculum*. Toronto, Canada: The Council of Ministers of Education, Canada.
The Royal Society. (2010). *'State of the nation' report on 5–14 science and mathematics education*. London, UK: The Royal Society.
van der Akker, J. (1998). The science curriculum: Between ideals and outcomes. In B. J. Fraser & K. G. Tobin (Eds.), *International handbook of science education* (pp. 421–448). Dordrecht, The Netherlands: Kluwer.
Vedder-Weiss, D., & Fortus, D. (2012). Adolescents' declining motivation to learn science: A follow-up study. *Journal of Research in Science Teaching, 49*(9), 1057–1095.

Chapter 2
Models of Modelling

Abstract The widely recognised importance of models in scientific practice suggests that they should have an equally important role in science education. The meanings attached to the word 'model' in the philosophical literature and the psychological literature leads to two canonical interpretations: the 'models as representations' view and the 'models as epistemic artefacts' view. The latter is argued as being the more educationally valuable of the two. The central role of analogy, thought experimentation, and argumentation, in the creation and validation of models – the act of modelling – is explored against a background of the philosophical and psychological literature on the theme. Ways in which models have been developed in educational contexts are then explored, leading to the presentation of the new version of the 'Model of Modelling' that is the basis of this book.

Introduction

In Chap. 1 we argued that, if science education is to address the challenges it currently faces, then the knowledge and skills of modelling have major roles to play. A wide-ranging discussion of the nature of modelling must be undertaken before its role in science education can be discussed, this leading to the identification of other major ideas on which it relies.

The first issue to be encountered is one of terminology, even of spelling. When searching Google for *modelling* (and *modeling*), one gets 306.100.000 results.[1] Taking into account the existence of several ordinary meanings for the word *model* (see the Preface), this number has no meaning in the current context of science education. However, when searching for *scientific modelling/modeling*, one gets 192.100 results. In a century in which the role played by the Internet in people's life is unquestionable, this datum also supports those who believe that modelling is important in science.

[1] In January 2013.

J.K. Gilbert, R. Justi, *Modelling-based Teaching in Science Education*, Models and Modeling in Science Education 9, DOI 10.1007/978-3-319-29039-3_2

By turning to a more specific context – that of those who were recognised as eminent scientists – one might be curious about the result of the same search of the Nobel Prize website. Here one gets 1.208 results, being 673 in Chemistry, Physics and Medicine, and 269 in Economics. Among the results in Chemistry, some are related to James D. Watson, laureate in 1962 for his work on the proposition of the structure of DNA and, undoubtedly, one of the most important scientists in the twentieth century. He is one of the scientists who have always recognised the role played by modelling in their work. In his most famous book – The Double Helix, (Watson, 1968) – he clearly shows not only the long and dynamic process of building concrete/material models, but also their manipulation and use in supporting the production of explanations, making predictions, and communicating ideas, that were absolutely crucial in the development of knowledge concerning the structure of DNA.

Some major questions emerge from such information: What is modelling? How is modelling performed in specific sciences? What is the meaning of 'model'? How do models and modelling participate effectively in the production of scientific knowledge? The answers to them can be derived from particular areas of knowledge (for example Philosophy of Science and Cognitive Psychology), and comprise an intricate set of ideas on which distinct proposals and actions in respect of science education can be based.

In this chapter, we discuss such questions in order to characterise a framework within which other ideas can be discussed in this book. Assuming the existence of complex relationships between them, we opted to start the discussion from the noun (*model*) rather than the verb (*modelling*) due to the apparent simplicity of the former and to its implications for the subsequent discussion about the latter.

Models

The nature of models has been explored extensively by both psychologists and philosophers. Taking these in turn:

Some Ideas from Psychology

Psychologists have also been discussing the meaning and role of models. According to Nersessian (2008), Kenneth Craik discussed the role of thought experiments involving mental models in human reasoning in a book published as long ago as 1943 (Craik, 1943). By taking into account the predictive power of thought and the ability of humans to mentally consider both real and imagined situations, he assumed that models were structural, behavioural, or functional analogues to real-world phenomena. However, as such ideas were published when the behaviourist approach dominated psychology, they received little attention. The second edition of his book

(Craik, 1967) had more impact on cognitive psychology, resulting in the development of research on the nature and function of mental models. In the 1980s, for instance, the publication of two influencing books both called "Mental Models" (Gentner & Stevens, 1983; Johnson-Laird, 1983) disseminated the research that was being conducted to that date.

Many years after Craik had come up with the idea that one reasons by using internal mental models of the world, Johnson-Laird (1980, 1983) recognised its originality when making his own contributions to the area. He assumed that a mental model is a structural analogue of a real world or imaginary entity that, as such, can be useful in providing explanations. Moreover, he clearly distinguishes mental models from 'pictures in the mind' (on the basis that the former do not need to possess pictorial attributes but the latter do) and from propositional representations (that is, a true or false description of something whose structure is not analogous to another structure). It seems he viewed mental models as iconic in nature (Nersessian, 2008).

In all major psychological studies conducted from the 1980s onwards, mental models are depicted as being internal representations of objects, events, or processes that have similar relation-structure to what is represented. In other words, they are structural analogues of what is being represented, constructed in the mind to reason with (Nersessian, 2002; Thagard, 2010; Vosniadou, 2002). This means that, for psychologists, mental models are not mental images, although in some cases they recognise that it is possible to associate an image with a given mental model. Moreover, psychologists assume that models enable individuals to explain and make predictions about phenomena, as well as to solve problems involving them. Cognitive psychologists have henceforward been mainly interested in investigating questions concerning the construction and use of such mental models.

Although the philosophical and psychological approaches were clearly distinct from each other, they share some ideas, as discussed next from a more detailed presentation of important philosophical discussions.

Some Ideas from Philosophy

Until around the 1970s models were discussed from the *syntactic view*, being defined in terms of theories. Under this logical positivist view of science, a theory is a syntactic structure involving a set of axioms, that is, it is a linguistic entity, and the role of models is to provide the conditions under which the axioms can be said to be true (Hartmann, 2008; Knuuttila, 2005a; Morrison, 2007). In other words, models are just a system of semantic rules that interpret the formal and abstract mathematical calculus (Frigg & Hartmann, 2009; Portides, 2011). Assuming the prevalence of such ideas, it is not surprising that Craik's ideas (see above) received almost no attention in philosophy.

From the 1960s onwards the *semantic* or *structuralist* view became the dominant one in use until very recent years. For their main proponents (da Costa & French, 2003; Suppe, 1989; van Fraassen, 1980), theories are non-linguistic entities defined in terms of models. Thus, a theory is a class of models or, looked at from the opposite direction, models are constitutive parts of theories. There is a general agreement among semantic philosophers that models are *representations* of reality. For them, reality consists of a fixed set of discrete objects ready to be represented, and models give us knowledge by providing that representation. This implies that the representative power of a model is associated with the degree of its validity, its truth. The problem is that these authors diverge on the actual nature of the representational relation between the model and the reality it represents (Portides, 2005; Suárez, 2003).

The philosophers who support the mathematical notion of models understand such a relation in terms of *isomorphism* (Suppe, 1989; van Fraassen, 1980) or partial isomorphism (da Costa & French, 2003). Isomorphism is a kind of mapping in which a given model represents its target system if their mathematical structures are correspondent to each other, that is, if there is a

> one-to-one function that maps all the elements in the domain of one structure onto the elements in the other structure's domain and vice-versa, while preserving the relations defined in each structure (Suárez, 2003, p. 228).

Some good examples in which the representation is based on isomorphic relations are provided by graphs and maps.

On the other hand, the relation between a model and the reality it represents can also be understood in terms of *similarity* (Giere, 1988).[2] Similarity is a vague notion associated with resemblance, implying that a model has similar properties to parts of a real-world phenomenon (that may be related to its visual appearance). Similarity almost always characterises the relation between concrete physical representations and the objects they represent. However, from the similarity point of view, it would not be possible to say that a mathematical equation written down on a piece of paper is a model, since it is not similar to any part of the phenomenon it represents (Suárez, 2003).

Another idiosyncrasy of different semantic views relates to the elements involved in the representation relation. Initially, semantic philosophers have assumed a representation to be a *dyadic* relation involving only the real system and the model, that is, assuming the existence of a straightforward relation between them. So, the key point would be the knowledge of the structure of the model and that of the thing being represented. More recently, some philosophers have recognised the role of representation producers and users in this relationship (Bailer-Jones, 1999; Giere, 2004; Morrison & Morgan, 1999; Suárez, 2003). Such a *triadic* relation was clearly expressed, for instance, when Giere (2010) wrote that a representation "cannot just be a matter of similarity between a model and the thing modeled" (p. 274). This means that an agent (A) produces or uses a model (M) to represent a part of the world (W) for some purpose. For Giere, the role of the agent is to specify both which similarities are intended and for what purpose. Thus, a model does not represent by itself: the representation occurs when someone uses a model (Giere, 2004). The triadic relation has two relevant implications. First, distinct models focusing on different aspects of a given reality can be produced; second, different agents can use

[2] Although it seems that, in more recent publications (for instance, Giere, 1999), he had changed his view by defining models as 'abstract entities' – something closer to the psychological mental models.

the same model for different purposes. This means a clear change in the focus from the notion of 'model of something' (or of 'how models represent the world') to the one of 'model for something' (or of 'how models are used to represent the world') (Knuuttila, 2005a).

Independently of the particularities of their views, all semantic philosophers base their views on the notion of *representation*. But, what does *representation* mean? The word representation originates from the word *raepresentare* used by the Romans as meaning 'to make present', 'to present again or to show/depict' (originally in the sense of "the embodiment of an abstraction in an object" (Pitkin, 1992, p. 3)). Another meaning of representation – human beings acting for others – began to emerge in Latin around the thirteenth century. More recently, representation has also become to mean 'to stand for'.

Such basic meanings have not changed much. They are found in ordinary dictionaries, have been discussed by philosophers, and have been used by scientists, students, teachers, and the general public. In the current context, the existence of such different meanings results in the parallel existence of different meanings for 'model' (Knuuttila & Boon, 2011). For instance, if someone assumes the meaning of *representation as showing*, he/she may be thinking from either an isomorphic or a similarity perspective. This means that such a person may assume models to be copies (even without this word being used) or as depictions with different levels of resemblance to the original. In both cases, that person would not identify, for instance, a graph where speed and time data were plotted as a model. On the other hand, *representation as standing for* implies in the presence of something that is being substituted for the entity that is being represented in the sense that allows the study and the drawing of conclusions about it. In this way, models could be used to learn about the world. So, someone with this view would not have difficulty in identifying the previous mentioned graph as a model, since it is not isomorphic with the phenomenon but can clearly be used to study it.

These multiple meanings of representation, as well as the absence of discussions about them among the semantic philosophers, have supported a criticism that has grown over the last decade. Some relevant foci of criticism have been:

- The idea that models are representations of something implies that we know enough about that something (that is, about its structure and/or other features) to be able to identify the content of its representation, as well as the way to represent it. This would strongly reduce the role of modelling in science, since models would have only a communicative role rather than an investigative one. In other words, scientific knowledge would be assumed to be already in existence, whilst modelling would not be a creative and complex practice that underpins knowledge building (Knuuttila, 2005a; Knuuttila & Boon, 2011; Morrison & Morgan, 1999; Murad, 2011; Portides, 2011).
- In some areas of science, it is not possible to identify what is being represented. The most relevant example is artificial intelligence, an area in which new realities are created (Knuuttila, 2005a). This implies that models do not only represent objects, events, or processes of the real world, but also data (Knuuttila & Boon,

2011) and ideas that support inferences – an aspect that is crucial in the production of scientific knowledge.
- The relationship between models and theories cannot be viewed as constitutive. Models are autonomous agents, that is, they are partly independent of both theories and the real world. This is so because models are not completely derived either from data or from theory. When they are produced, both data and theory are involved as well as are other elements (for example: analogies, mathematical equations). Such an autonomy justifies the use of models as investigative tools in scientific practices, as vehicles for learning about the world (Frigg & Hartmann, 2009; Morrison & Morgan, 1999).

Such criticisms lead to the proposal of different views on models. Among those, we emphasise two on the basis of their current or possible future implications for science education. They have not been clearly named in the literature, but are named here as the *'mediation'* view (proposed basically in Morrison & Morgan, 1999) and the *'artefactual'* view (proposed by Knuuttila, 2005a, 2011).

From the idea that models are partly independent from both reality and theory, as previously emphasised by Cartwright (1983), Morrison and Morgan (1999) propose that they function autonomously, that is, that they mediate between reality and theories and can be used for different purposes. They also characterise models as investigative tools which, for them, means that they "involve some form of representation" (p. 11) since models may represent some aspects of the reality, some aspects of the theories about reality, or even both. However, they affirm that they have not used the notion of 'representing' as synonymous to 'a kind of mirroring', but as

> a kind of rendering – a partial representation that either abstracts from, or translates into another form, the real nature of the system or a theory, or one that is capable of embodying only a portion of a system (Morrison & Morgan, 1999, p. 27).

This justifies the simultaneous existence of different models of something, as well as the use of each of them for different purposes and in specific contexts (from measurement activities to the design, production, and application of both theories and new technologies) (Morrison, 2011). Moreover, they emphasise that the value of a model is not related to the extent to which it is an accurate representation of something (as accepted in the semantic view), but to their performance in specific contexts (for instance, their explanatory power in these contexts). In another publication, Morrison (2007) also emphasises that a certain degree of representational inaccuracy is one of the essential characteristics of a model because it may be required for a model to be used for a particular purpose.

Another particular context that is relevant to scientific enquiry is that of investigating systems that are inaccessible for any reason through the use of simulations. Morrison and Morgan (1999) highlight the use of models in simulations of behaviours or phenomena, even when we do not know their real characteristics. This seems to imply that they see models as also representing ideas and having predictive power – which clearly characterises them as active agents in the production of scientific knowledge, rather than as subordinate to data and theory. Similar ideas related to the roles of models in knowledge building are now acknowledged by cognitive psychologists when explaining how scientists think (Nersessian, 2002;

Vosniadou, 2002), although all of them relate such roles to the representative nature of models.

From an in-depth analysis of both the views of the main semantic philosophers and the ideas expressed by Morrison and Morgan (1999), the Finnish philosopher Tarja Knuuttila (2005a) expanded the later ideas and proposed a novel and well-based view of models in science. According to her, the consequence of turning the model-as-mediator approach into an effective alternative to the model-as-representation approach is

> to free models from the theory data framework still present in Morrison and Morgan (1999) and to interpret models materially, thus granting them an individual status as *epistemic artefacts* (Knuuttila, 2005a, p. 48, our emphasis).

Therefore, models are viewed as human made artefacts materialised in some way, which results in them having "many other epistemic functions besides that of representing the world" (p. 18).

This conveys a distinct status to models as independent agents. Whilst Morrison and Morgan (1999) and Boumans (1999) propose this independence in terms of the theory-data framework, Knuuttila (2005a, 2005b, 2011), followed by Weisberg (2007), proposes that models are independent of real-world target systems. According to her, this is essential if we aim at understanding how models enable us to learn from producing and using them.

By viewing models as artefacts rather than as representations, Knuuttila assumes that they can support our imagination on how objects can be, behave, and interact with each other. In other words, being artefacts, models can be used in many scientific practices and in distinct ways. Therefore, as previously emphasised by Morrison and Morgan (1999), we could generate knowledge and learn from models by producing and using (manipulating) models, that is, by modelling (Knuuttila & Boon, 2011). In such a complex process, one of the multiple epistemic practices performed by models may be that of representation, but this can be understood as a mediative and creative activity involving objects, process and ideas, sometimes being imagined during the process. Other and more essential practices are the productive ones, that is, those related to specific purposes or to the scientific questions in relation to which they are produced: like making simplifications and idealisations; conceptualising imagined (or non-directly observable) objects or processes; supporting arguments, explanations, and predictions. The performance of a model in terms of such epistemic practices could, then, be used to assess its success (rather than the accuracy of the representation, as claimed by the semantic philosophers) (Knuuttila, 2011).

Some details of this novel focus on material artefacts have to be clearly understood. First, 'material' does not mean a restrictive view on concrete objects. It rather emphasises the need of a model to be "expressed by external representational means" (Knuuttila & Boon, 2011, p. 315) that favour its manipulability. As models can be expressed in different modes (like equations, symbols, graphs, dynamic images), scientific reasoning can be performed in distinct (sometimes abstract) ways. Thus, material artefacts are not opposite to theoretical or abstract things. As stressed by Knuuttila,

> as parts of collective human environments, material things are already endowed with interpretations, meanings and knowledge concerning them (Knuuttila, 2005b, p. 1267),

This association of the idea of artefacts with that of models

> stresses their multifunctionality in science and suggests that the links between models and reality are more complex that appears when we focus only on representation (Knuuttila, 2005b, p. 1267).

This is so because (i) models are linked to our knowledge about the world through the questions addressed in their production; and (ii) most knowledge produced during modelling results from different kinds of inferences about the entity that is being modelled that are drawn during modelling. In sum, models function as external artefacts to support thinking, whilst their construction and manipulation support their performance of several epistemic functions.

In this book, we consider the notion of 'model' as characterised in the *'artefactual'* view. We do so because there is a complete coherence between its main arguments and the ideas we have been developing in the last decade from our investigations on modelling-based teaching conducted in regular science classes. In previous publications (for instance Gilbert, 1993; Gilbert & Boulter, 2000; Justi, 2006; Justi & Gilbert, 1999, 2003) we had generally assumed a semantic view (mainly by characterising models as partial representations). However, when we have involved students in modelling activities, it became clear that they perform such activities and build knowledge from such an experience by thinking about, and working with, models as epistemic artefacts. Moreover, our own initial proposal for modelling (Justi & Gilbert, 2002) explicitly emphasises both the relevance of the materiality of models and their multiple purposes and uses. More details about these (currently realised) similarities are discussed in the following section, as well as in other chapters of this book.

Modelling

Modelling has been understood as a process of producing, or building, models (a meaning that appears in most ordinary dictionaries), or as a broader (or more detailed) process that also includes the use, or manipulation, of models. In both cases, it is a complex process that has attracted the attention of many scholars interested in how knowledge is produced. As in the case of models, modelling can also be approached from philosophical or psychological perspectives. However, here the differences between their contributions are huge due to the nature of the questions that each of them addresses. Philosophers are generally concern with the relationships, or logical operations, between models and theories, or between models and target systems, depending on their understanding of the nature of 'model'. On the other hand, psychologists are interested on how mental models are created, manipulated, evaluated, and used when one thinks about a target system (Nersessian, 2008). Both of them contribute to a composite wide-ranging view on modelling.

Philosophical Contributions

By defining models as representations of reality that constitute theories, the views of semantic philosophers about modelling depend on the specific relation that they propose between models and realities. Those who assume that models are isomorphic mathematical structure-types view the production of models as the result of either the use of mathematical language (Suppe, 1989; van Fraassen, 1980), that is, in a more abstract way, or from a set of theoretical predicates (da Costa & French, 2003). On the other hand, those who defend models as being similar to their targets are mainly focused on building models that represent as many features of the target system as possible. Depending on the features to be represented, models would be built from the use of different modes of representation (Giere, 2004). For instance, some of the features of a pendulum could be represented in a drawing or 2D diagram, whilst those related to their movement would be better represented by building a concrete 3D model.

However, such distinctions are, in some sense, superficial ones, since all semantic philosophers were concerned only with the production of the representations, with the underlying structure or feature that had to be represented. As far as we know,[3] they have neither focused their attention on modelling nor recognised it as a complex process that might be explored – which is in accordance with their view of models as constituents of theories (the main scientific entities).

More recently, under the mediator and the artefactual views, models were claimed to be independent of theories. By assuming models' independent identity, modelling has also started to be reconsidered. This was done not by trying to explain details of the process, but by recognising its complexity (due to the diversity of elements involved, like analogies, mathematical techniques, empirical data, previous theoretical views, etc.) and the resulting impossibility of fixed rules for model building (Boumans, 1999; Morrison & Morgan, 1999). Morrison and Morgan also emphasise two additional ideas related to the lack of a recipe for modelling. First, modelling may be viewed as an art rather than as a mechanical procedure, which is an important idea because it recognises the essential role of creativity and imagination in the process. Second, we learn not only by building or seeking to comprehend a given model, but also by using it. As the notion of representation is still present in the mediator view, learning would also be related to the use and interpretation of different modes of representation when building and using models. Such ideas are shared by others, such as Frigg (2010).

On the other hand, by going beyond the notion of models as representations (although recognising that to represent may be one of the functions of models), other philosophers highlight distinct characteristics or functions of modelling. Peschard (2011), for instance, discusses the generative constructive nature of

---

[3] As we are not philosophers, we had not exhaustively studied the whole work produced by the semantic philosophers. Rather, we studied only their main works in order to understand their general ideas on models and modelling, which would be important to fulfil our aims for this book.

modelling, that is, the fact that modelling can support the construction of other models as well as the generation of other target systems. This implies modelling as being a continuous investigative process. Additionally, the shift of focus from representation to epistemic artefacts proposed by Knuuttila also sheds light on the activity of modelling as a complex scientific practice. According to her, modelling cannot be separated from the construction of new phenomena and knowledge – processes by means of which models acquire their epistemic values (Knuuttila & Boon, 2011). So, modelling is guided not only by the mode of representation used to produce the epistemic artefact (which would be only one step in the process), but also mainly by the epistemic purpose addressed in our thoughts during the process. This also implies that modelling is a developing practice in which one aspect (the use of any mode of representation to materialise the model) is followed by the acquisition of its epistemic value. On the contrary,

> the concrete embodiment of a model (whether symbolically, iconically or three-dimensionally rendered) draws together and integrates, *in each stage of its development*, the various empirical, theoretical, and conceptual dimensions of model construction (Knuuttila & Boon, 2011, p. 316, our emphasis).

It is worth noticing that they explicitly mentioned the occurrence of distinct stages in the process. This is clearly shown in Knuuttila and Boon's paper, when they analyse the history of the Carnot Cycle model and demonstrate how a model can simultaneously provide answers to previous questions and generate new questions, that is, how it can also become a tool for its own development. We interpret their discussion as evidence that modelling is a cyclical process guided by problems to be solved from several epistemic practices rather than by representational purposes. The main focuses of Knuuttila and Boon's discussion are exactly on the key role of conceptualization in modelling, and on "how modelling consists of a co-construction of different mutually developing elements that are drawn together by the model" (Knuuttila & Boon, 2011, p. 330).

Psychological Contributions

As previously mentioned, Craik was the first psychologist who hypothesised that humans reason through modelling. By drawing an analogy between a machine and the human brain, that is, by assuming that they share similar relation structures, he proposed that reasoning occurred through simulations involving mental models. For him, this occurred not only because "some people think more easily and fruitfully in terms of mechanism and things that can be visualised" but also because "a model gives a general indication of the kind of thing that can be mechanically achieved" (Craik, 1967, p. 110). So, his ideas on modelling were related to how a model functions since this may be the basis of explanations about the target of the model.

When Johnson-Laird proposed his psychological theory on mental models (1980, 1983), he explicitly demonstrated that, in his view, the logical components of modelling were only procedures for testing mental models, that is, for trying to "establish the falsity of a putative conclusion by destroying the model from which it derives" (Johnson-Laird, 1980, p. 83). For him, this would support individuals' understanding of the entity being represented, as well as the making of inferences and predictions about it. The use of this well-accepted principle of formal logic seems to have contributed to the dissemination of his ideas among the communities both of psychologists and of science educators. Therefore, most (if not all) of these communities recognise (with more or less emphasis) the existence of a test phase in modelling.

Another cognitive psychologist who has based her work on the conceptualisation of mental model as a structural analog of a real world or imaginary entity produced by human minds to think with is Nancy Nersessian. In the last two decades she has published a significant contribution to the discussion on the production of scientific knowledge, conceptual change, and human cognition from the psychological point of view (exemplified by Nersessian, 1992a, 1992b, 1999, 2002), organised in the book Creating Scientific Concepts (Nersessian, 2008). From a thorough analysis of the historical scientific practices as performed by James Maxwell, she discusses how mental models underlie the creative reasoning of scientists.

By assuming the intrinsic relationship between internal and external representations, that is, mental modelling and external physical models, she acknowledges that model-based reasoning "is not an ancillary aid to reasoning carried out by logical manipulation of propositional representations" but that it is performed "to satisfy constraints deriving from the target and source domains, and those that might emerge from the model itself" (Nersessian, 2008, p. 184) in a dynamic process. She also considers that this kind of reasoning involves bootstrapping, that is, "consists of cycles of construction, simulation, evaluation, and adaptation of models that serve as interim interpretations of the target problem" (Nersessian, 2008, p. 184). This type of reasoning involves the creation of an initial mental model from the interplay of the question that guides the problem being faced, some possible ideas related to it, and the selection of an analogical source domain. The production of this initial mental model is followed by its test from the analysis of the mapping relations between the target and the model in order to identify if and how they allow, for instance, the production of satisfactory explanations and predictions from the use of the model. During this process, models are modified in different respects and over a period of time until the answers provided by the model to the problem are accepted by the relevant community. This view relates coherently to that discussed by Knuuttila and Boon (2011).

Even more important than this general framework for model-based reasoning is the detailed analysis that Nersessian provides of how mental models are created, manipulated, evaluated, and adapted in problem-solving situations, that is, on how modelling is performed. She shows that modelling is essentially imaginative because it is based on three epistemic, creative, and inextricably intertwined practices: the use of analogies, imagistic representations, and thought experiments. Although other psychologists discuss some of these practices, she presents them as constituting a coherent whole, which justifies our option for focusing on her ideas.

Analogies are essential because models are produced and tested on the basis of relational comparisons. This is so because analogies are generative in such reasoning processes (through the main sub-processes discussed in the psychological literature on analogies: retrieval, mapping and transfer), as well as in the building of the model representations (an aspect that, according to Nersessian (2008), has not been addressed). It is worth noticing that, in science, analogical domains are always unknown in advance (Nersessian, 2002), that is, analogies are drawn to serve a creative function. Only later in the process they may assume a communicative role.

The emphasis on imagistic representations derives from the existence of the close relationship between internal and external representations. To classify an internal representation as imagistic is not to say that its format is picture-like or that there are overt resemblances between it and an external representation. The essential point is the existence of correspondences between interpretations of the elements of the internal representation and aspects of the external one. Therefore, the external representations may include not only drawings, diagrams or any other kind of two-dimensional image, but also gestures since all of them can support the organisation of cognitive activity during modelling (including the performance of simulations), the communication of mental models within and outside the scientific community, and the subsequent discussions of distinct models. Moreover, it is possible to transform one representation into another from both the knowledge of particularities of the codes of representation used in each of them and from the consideration of how each representation relates to particular aspects of the problem in question. Therefore, the use of visual external representations provides the materiality required by the philosophical artefactual view of models.

Finally, thought experiments involve mental simulations conducted with the mental model in order to investigate the outcomes of possible specific manipulations, that is, they occur as a result of thinking about the question 'What would happen if…?' This kind of reasoning clearly requires the internal selection, re-experimentation, imagination, visualisation, and anticipation of characteristics and behaviours of the entities being modelled, as well as the integration of previous knowledge in order to produce inferential transformations of the current mental model (Nersessian, 2002, 2008). In this way, thought experiments are crucial when revising and evaluating a model.

In a recent publication, Chandrasekharan, Nersessian and Subramanian (2012) discuss how the role of thought experiments has been decreasing in some areas of science due to the use of computational modelling. They exemplify how this replacement has been occurring in bio-sciences and engineering fields mainly due to the deeper insights into problems provided by building computational models, and to the sophisticated simulation operations that can be supported by computational models. However, independently of this novel focus of attention, they continue to recognise the importance of the simulative reasoning with mental models that is the essence of the role played by thought experiments. Therefore, this comprehensive psychological approach to modelling sheds light on the intellectual work carried out by scientists. As it is very well characterised by Chandrasekharan et al. (2012), this is not an interpretative but rather an interrogative process, for it provides a relevant support for our understanding of the creative enterprise of science.

There is also another relevant epistemic practice associated with modelling: argumentation. This is so because, in the scientific context, argumentation can be understood as a process of justifying claims that occurs from the coordination of data and scientific theories or previous knowledge. As emphasised by Deanna Kuhn (1991), an argument by an individual is reasoning based on the initial production of an affirmative statement, followed by a justification, and in which opposite views are analysed. However, the phrase 'opposite views' does not necessarily imply a discussion between two or more people. Argumentation can also occurs in one's mind, when considering the existence of rival ideas concerning a model, or even rival models. When arguments are focused on models, their claims and justifications can be based on the analyses of the extent to which the models fulfil their purposes: this also plays a persuasive role within the community of scientists.

Models and Modelling in Science Education

The inclusion of models and modelling as an area of investigation in science education initially occurred during the 1980s. This resulted in the publication of a general paper on the use of models in science teaching (Gilbert & Osborne, 1980), as well as some studies of students' mental models of specific scientific topics by, for instance, Tiberghien (1994) and Vosniadou and Brewer (1992, 1994). Those studies emerged among the huge literature on students' ideas that dominated science education research at that time. Initially, it seems the word 'model' was used only to refer to scientific models taught to students. For instance, in the introductory chapter of an influenced book, Driver, Guesne, and Tiberghien (1985, p. 2) ask "Do the ideas that children possess represent coherent models of the phenomena that are frequently presented in classroom settings?". In the same chapter, they recognise that students' ideas, although stable, are often contradictory, which means that they do not have "a completely coherent model of the phenomena presented, at least in the scientists' sense of the word coherent" (p. 3–4). But the word 'model' was also used in the phrase "children's models" as synonymous with frameworks, or structure of ideas, sometimes being qualified with the adjective 'alternative' to distinguish students' from scientists' domains (Driver et al., 1985). At the same time, perhaps being influenced by the semantic philosophers and/or the cognitive psychologists who worked from the mental model perspective, some science educators started to adopt the concept of model as representation that could help in teaching (for instance, Harrison & Treagust, 1996; Ingham & Gilbert, 1991; Seddon & Moore, 1986). It was at the beginning of the 1990s that the meaning attributed by students and experts to the word model was first investigated (Grosslight, Unger, Jay, & Smith, 1991). Although we recognise the significant role that the research on models played in the growth of knowledge in science education, as the focus on this book is on modelling, we will not comment on it in detail.

On the other hand, modelling as a way of building knowledge is a notion that was initially mainly used in the context of science education by John Clement (Clement, 1989; Clement & Brown, 1989). By assuming that learning via model construction is a key process in science knowledge building, as well as recognising the lack of understanding about such a process at that time, he analysed a case study in which a scientist was asked to think aloud when solving a problem that was at the frontier of his personal knowledge (a situation that approximated to the generation of scientific knowledge). Prior to this, Clement had investigated how some philosophers thought about the mental processes involved in the formation of hypotheses in order to identify "some sort of model of the mental processes being used" (Clement, 1989, p. 345). The outcome was the proposition of the 'model construction cycle' (Fig. 2.1). This is a diagram that represents a cyclical process constituted by hypothesis generation, rational and empirical testing, and modification or rejection of a model. In this cycle, hypothesis generation constitutes a cycle sub-process in which the elements "make initial observation", "activate possible analogies and related model elements", and "construct initial model" are linked by double arrows in order to indicate the mutual interactions among them. Clement also emphasised the existence of a missing element: the influence of the subject's prior theoretical framework. According to him, it could affect the whole process, thus being difficult to depict in a 2D representation.

The analysis of Clement's detailed case study corroborated the model construction cycle and shows that the creative mental process can be divided into "the productive processes of generation and modification and the evaluative processes of empirical testing and rational evolution" (Clement, 1989, p. 369). Thus, when dis-

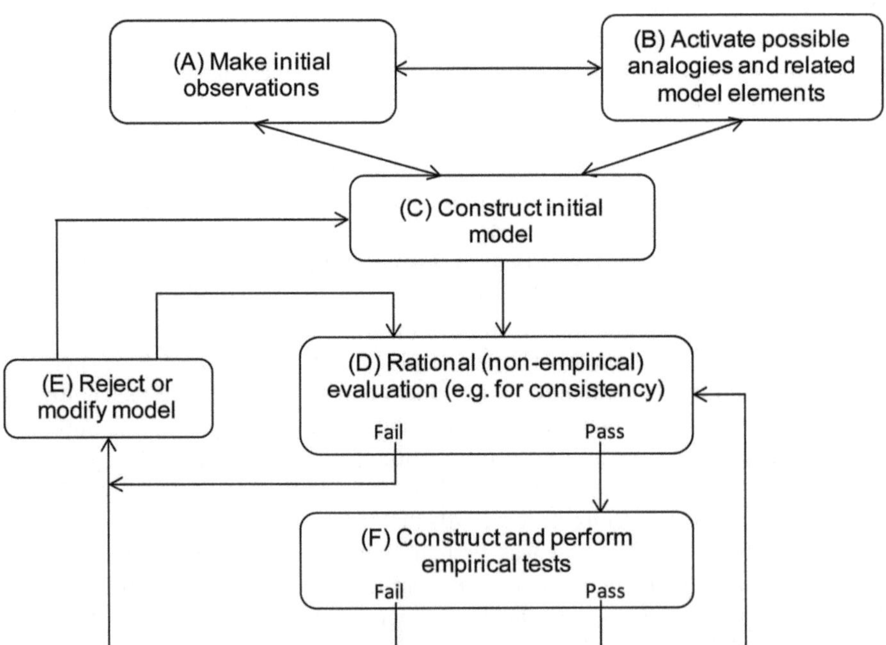

Fig. 2.1 Model construction cycle (Clement, 1989, p. 347)

cussing the implications of this proposal for science education, Clement suggested that the model construction cycle is a potential model with which to support the learning of scientific concepts if learning is viewed as students' constructing knowledge. In order to do so, the cycle has to guide teachers in designing instructional activities that could support students' experience of the whole process.

Before Clements' publications, other ideas about modelling in the context of physics education had been published (Halloun & Hestenes, 1987; Hestenes, 1987). However, such ideas were focused on mathematical modelling, and were mainly published in a journal whose target audience is college and university physics teachers and students rather than the science education research community. A paper was published in a science education research journal only some years later (Halloun, 1996). This seems to explain why the science education research community as a whole has not made many references to their work. Hestenes and Halloun's ideas are based on the semantic view of models. For them, a model is a representation of a real thing and "models in physics are mathematical models, which is to say that physical properties are represented by quantitative variables in the models" (Hestenes, 1987, p. 442). From these ideas, Hestenes proposes that modelling is a process constituted of four stages:

- *description* of the objects, properties and phenomena to be modelled, as well as physical interpretation to all variables involved, selection of the theory to be applied, and definition of the kind of model to be built;
- *formulation* of equations for the phenomenon from the use of adequate physical laws;
- *ramification*, a mathematical process in which the equations are solved, and the results are represented (analytically or graphically) in order to favour their analysis;
- *validation*, that is, the empirical evaluation of the ramified model.

Hestenes (1987) proposed the use of this modelling approach in the design of physics instruction. When teaching from such a 'model-centred instructional strategy', the teacher would have to show that problems are solved by the development of models, to explain the relevance of each stage during its discussion (thus teaching the organization of scientific knowledge), and to encourage students to employ the modelling strategy when trying to understand the textbook (Hestenes, 1987) or experimental activities (Halloun, 1996). In short, this approach is firmly focused on the learning of established models.

Our own view on modelling was proposed more than a decade after the Clement view, that certainly inspired and informed our ideas. In a paper in which we discuss diverse teachers'[4] views on modelling (Justi & Gilbert, 2002), we presented, for the first time, the *Model of Modelling*, a diagrammatic representation that depicts modelling as a process in which a mental model is produced and expressed in any mode of representation, the expressed model is tested (being modified or rejected), and the resulted model has its scope and limitations considered. Figure 2.2 presents a modi-

[4] From primary school to university level.

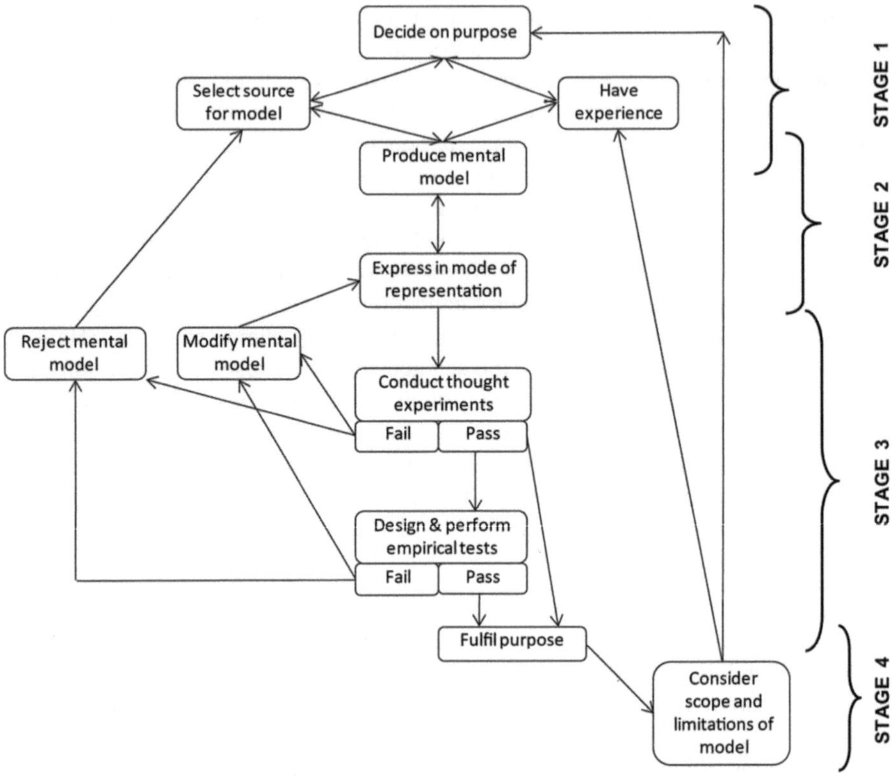

Fig. 2.2 Modified version of the Model of Modelling diagram (Originally published in Justi & Gilbert, 2002, p. 371)

fied version of the original representation, in which we add the identification of each of the stages that are detailed explained next.

At that time, we also viewed a model as a representation of a given entity (object, event, process or idea) produced for a specific purpose (Gilbert, Boulter, & Rutherford, 1998). However, after a decade of investigating students' learning from this perspective, studying models and modelling from philosophical and psychological points of view, and critically rethinking our ideas, we present our current view on modelling, which integrates the whole set of ideas developed during and since this time. It is slightly different from the original one and, in our view, can be more strongly justified.

First of all, as mentioned earlier, we now accept the artefactual view of models, that is, we understand models as epistemic artefacts, the purposes of which are related to many scientific practices like simplifying, explaining, abstracting, arguing, predicting, representing, designing experiments and/or other models, etc. Both the diversity of such practices and the role they play in science justify the relevance of models in the development of scientific knowledge. Second, we consider modelling as a complex cyclical and non-linear process of knowledge construction basically comprised of the same four stages identified in the original proposal, but named, related to each other, and justified (as is explained next). Moreover, it is not predetermined, that is, in general there is no specific order of occurrence of the stages or the sub-processes in each stage.

In a given context, the *creation* of a fresh proto-model initially involves the integration of three elements: *purpose(s)*, *experience(s)*, and *source(s)*. In order to be involved in the process, one – the modeller – defines an initial purpose(s) for the model or understands the purpose(s) defined by someone else (which is very common nowadays, when scientists work in big groups in which individuals collaborate with each other). The experience(s) to support the creation of the proto-model can be acquired from revision of one's cognitive structure (that is, personal previous knowledge), study of the related literature (that is, knowledge previously developed by someone else), and/or analysis of empirical data. As purposes may originate from creative view of experiences, there is no specific order of occurrence of the elements of this stage. Initially a proto-model is created from the definition of the initial purpose. Then, it can occur at any time as a new purpose emerges or the original one is modified. This means that this stage can be associated with *creation* or *development* of a proto-model. Finally, the source of the proto-model may be an analogy or a mathematical tool used to establish relationships between elements of the experience, thus supporting its creation or development.

One feature that distinguishes our view on modelling from that proposed by John Clement is that the *creation of the proto-model* occurs in one's mind, that is, it corresponds to what is generally called a 'mental model'. However, we would rather call it proto-model in order to establish a cohesive relationship between this nomenclature and the idea of models as artefacts. This requires the existence of another stage in modelling: the *expression of the proto-model* in any of the modes of representation or combination of them, thus resulting in the creation of a model *per se*. Therefore, what is expressed is not a representation of the target entity of the model (that is, a model is not itself a representation of the reality). It rather represents the proto-model, the entity created in one's mind to act as an epistemic tool (or to address a given purpose). In this sense, 'representation' is being used here meaning 'to stand for', that is, implying the presence of something that is substituted for the entity that is being represented: this allows the study and the drawing of conclusions about it.

The most common modes of representation are: two-dimensional (drawings, diagrams, maps, sketches, etc.), three-dimensional (concrete, either in different scales or analogical), virtual (including simulations), gestural, mathematical, and verbal. This means that we view artefacts not only as concrete objects, but as whatever external representations that can be manipulated in different and suitable ways – something that was recognised by Knuuttila in a recent publication (Knuuttila, 2011).

The selection of the mode of representation to be used is a key action guided mainly by: the purpose(s) addressed in the process, the nature of the elements to be modelled (static or dynamic, concrete or abstract), the epistemic practices that will be conducted with the manipulation of the model (some of them may be limited by specific modes), and its target public. Concurrently with or after the selection of the mode of representation, the modeller also defines the codes of representation, that is, the meaning of specific details of the resulted artefact. For instance, in a concrete ball-and-sick model of a chemical substance, it is necessary to specify that the balls represent the atoms that constitute the substance, that the sticks represent covalent bonds, and that distinct colours for balls represents specific elements. It is worth emphasising that the expressed model does not entirely coincide with the proto-model that originates it. This is so because there are limitations to the modes of representation that can be used, for, on the one hand, these can only partially express the proto-model, and, on the other hand, some of them require specific and sophisticated skills or knowledge that may not be available to the modeller.

This stage of expressing the internal representation in an external form has perhaps been overlooked in other proposals for modelling in science. However, we view it as essential because models can only perform the epistemic practices for which they are produced if we interact with them. In other words, we acknowledge that cognition is distributed, that is, that the expressed models are not only memory aids (Giere, 2002; Magnani, 2002), that cognition is not only in the human mind or in the world (the reality itself and the artefacts built in relation to it), "but 'in the system' such that an individual's mental activities comprise interactions with other material and informational systems (including other humans)" (Nersessian, 2008, p. 117). Such interaction facilitates not only the selection of the mode of representation to be used in this stage of the process, but also a possible change in the purposes of the model, or a search for new experiences to be considered in its development, and the conduct of a later stage: the *test* of the model.

The *test of the model* can be empirical and/or mental, that is, it can involve designing and conducting empirical and/or thought experiments. It depends on the entity being modelled and on the resources or conditions available with which to conduct the test. Moreover, this stage does not consist of a single test. In science, a model only becomes acceptable for that moment in time after achieving success in a series of tests. It is also very common that a thought experiment precedes an empirical one, when this is possible. If a model fails a test of either type, it has to be modified, which may imply the adding of new experiences or reinterpreting of previous ones (in the light of the outcomes of the test), or even in selecting a different source for the model (for instance, seeking an analogical relationship). The tests can also show that the model has basic flaws, resulting in its rejection (Justi, 2006).

Finally, the *evaluation of the model* involves the identification of the scope and limitations of the model, and occurs from attempts to use the model in different contexts. This is also a stage in which the modeller tries to convince others about the validity and usefulness of the model. It may be possible to identify a limitation that really restricts the use of the model. This results in either a return to the 'model of modelling' cycle for the model to be modified, or the acceptance that it can only be used in specific context(s).

The performance of each stage, as well as the making of all necessary decisions related to the definition of the steps to be addressed next during the process, are mainly guided by the cognitive processes previously identified by Nersessian

(1992a, 2002, 2008): the use of analogies, imagistic representations, and thought experiments, as well as by (internal or external) argumentation.

Analogical reasoning is important not only for selecting the source for the proto-model and testing the model (as previously emphasised by Nersessian). We view analogical reasoning as essential to establish relationships between different experiences (that is, as part of the creative process of integrating different elements in the creation of the proto-model); to support the selection of the modes of representation to be used; to express a given proto-model as an analogy; to form the basis of the design of a test; to support a simulation; to establish or analyse relationships between the use of the model in different contexts; and to convince others about the validity and utility of the model.

As imagistic representations are associated with the relationship between internal and external representations, they are also involved in the whole process. Such relationships can be established during the creation, expression, test, and evaluation of a model. This includes attempts to persuade others of the validity of the model. Although thought experiments are mainly performed when the model is being tested, they can also be used in the creation stage (when one is attempting to establish relationships between different types/items of information), in the expression stage (when one analyses the adequacy of a mode of representation for the expression of a given idea), and in the evaluation stage (when one both analyses the behaviour of the model in another context and tries to convince others of the utility of the model). Finally, argumentation permeates the whole process of discussion of the distinct ideas that occur in all the stages of modelling, since the adequate justification of claims is essential in order for meaning to be attributed to models, and to persuade others about their utility and validity.

From this description, it becomes evident that modelling is really a complex, dynamic, non-linear, and creative cyclical process. As so, each stage does not follow another one in a strict order, although sometimes a given action requires the occurrence of another specific action. Boumans (1999) draws an analogy that clearly illustrates such situations. According to him, "model building is like baking a cake without a recipe" (p. 67). We can develop the analogy further in order to convey a support of its understanding and to justify the presentation of a diagrammatic representation of the process.

When one lives in a country with a long and strong tradition of baking, a person has no difficulty in understanding that there is not a single recipe for a cake. In the U.K., for instance, a brief browsing in books on baking shows that there are more than 2500 different recipes for cakes. Such diversity is explained not only by the use of distinct ingredients (as emphasised by Boumans), but also by both the performance of specific different procedures and the desire to reach particular outcomes (for instance, in terms of softness of the cake). However, the baking of all cakes implies a following of some basic common steps (even if in different orders or to different extents). Likewise, although it is impossible to produce a recipe for modelling, it is possible to organise such basic stages in a way that could provide an overview of the process, thus helping those who are not familiar with it to understand its

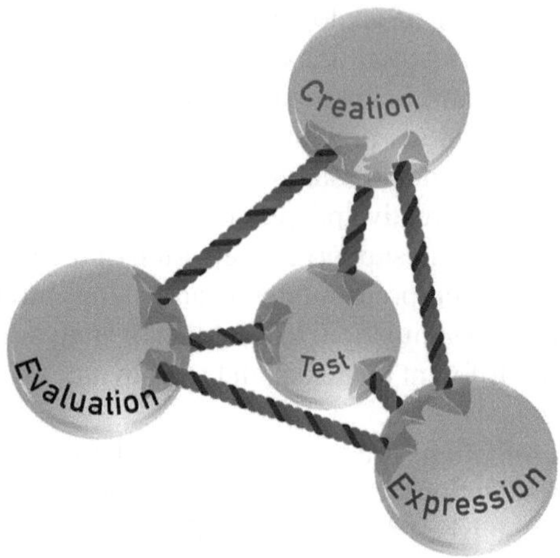

Fig. 2.3 The Model of Modelling v2

meaning. In the same way, it is possible to teach a child to bake from explaining both the basic procedures to be followed and the most common variations of them.

By taking this analogy as a scaffold, we present a two-dimensional artefact of our views of the nature of modelling in science. We name it Model of Modelling v2, since an original version was presented more than a decade ago (Justi & Gilbert, 2002). The Model of Modelling v2 (Fig. 2.3) shows a tetrahedron in which each of the main stages of the process is placed in a vertex. We opted for a tetrahedron because (i) it is a geometric shape in which all the vertices are equidistant from each other, and (ii) it can be rotated without changing the relationships between the vertices. Therefore, the use of this geometric shape emphasises three essential characteristics of the process: to be cyclic, non-linear, and not predetermined. Moreover, the edges of the tetrahedron are represented as a string made of four threads twisted together. Each of the four threads represents one of the cognitive processes involved in the whole modelling process: the analogical reasoning, the use of imagistic representations, thought experiments, and argumentation. This justifies the use of distinct colours for representing them. Figure 2.3 shows the strings not only connecting the spheres of each vertex, but also entering them in order to emphasise that the four cognitive processes occur all the time.

One of the advantages of the Model of Modelling v2 is that it can also depict the sub-processes involved in each of the stages (Fig. 2.4), as well as the elements constituents of each of them (Fig. 2.5). So, our main ideas on modelling are expressed in the set of Figs. 2.3, 2.4, and 2.5.

The Model of Modelling v2 was produced by having in mind that modelling is a scientific epistemic practice. Although we have used several visual elements in order to make the expression of our ideas clearer, being a two-dimensional artefact it has its limitations. For instance, it may be difficult for someone to visualise the tetrahedron in three dimensions, that is, to "view" that all the vertices are at the same distance to each other, that the angles are exactly the same (which means the

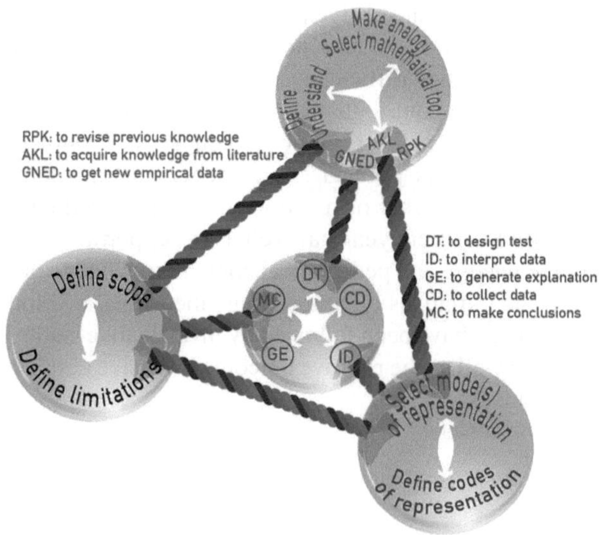

Fig. 2.4 The Model of Modelling v2 (With emphasis on the main sub-processes involved in each stage)

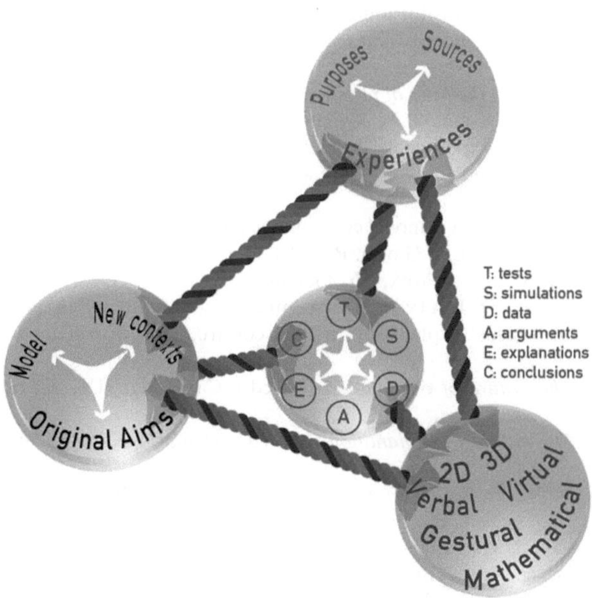

Fig. 2.5 The Model of Modelling v2 (With emphasis on the main elements that participate in each stage)

tetrahedron can be rotated and no relationship between the elements will be changed), and that the central sphere is behind the other three. But assuming that this and other possible minor difficulties may be overcome, it is likely that the Model of Modelling v2 can contribute to an enhancement of one's view on modelling. As such, it can be very helpful in science education. From a consideration of the aspects represented in Figs. 2.4 and 2.5, a teacher can plan both activities to be performed by students and his/her own actions in order to facilitate students learning. It has been so used in recent years, as well as in supporting the development of teachers' content knowledge and pedagogical content knowledge on modelling (see Chap. 11). Most of the initiatives of "translating" the Model of Modelling into science educational contexts have been followed by investigative studies. The nature of such investigations, as well as the positive outcomes we have obtained are discussed in other chapters of this book.

References

Bailer-Jones, D. M. (1999). Tracing the development of models in the philosophy of science. In L. Magnani, N. J. Nersessian, & P. Thagard (Eds.), *Model-based reasoning in scientific discovery* (pp. 23–40). New York, NY: Kluwer and Plenum.

Boumans, M. (1999). Built-in justifications. In M. S. Morgan & M. Morrison (Eds.), *Models as mediators: Perspectives on natural and social science* (pp. 66–96). Cambridge, MA: Cambridge University Press.

Cartwright, N. (1983). *How the laws of physics lie?* Oxford, UK: Clarendon.

Chandrasekharan, S., Nersessian, N. J., & Subramanian, V. (2012). Computational modeling: Is this the end of thought experiments in science? In M. Frappier, L. Meynell, & J. B. Brown (Eds.), *Thought experiments in philosophy, science, and the arts* (pp. 239–260). London, UK: Routledge.

Clement, J. J. (1989). Learning via model construction and criticism: Protocol evidence on sources of creativity in science. In J. A. Glover, R. R. Ronning, & C. R. Reynolds (Eds.), *Handbook of creativity* (pp. 341–381). New York, NY: Plenum.

Clement, J. J., & Brown, D. E. (1989). Overcoming misconceptions via analogical reasoning: Abstract transfer versus explanatory model construction. *Instructional Science, 18*(4), 237–261.

Craik, K. (1943). *The nature of explanation* (1st ed.). Cambridge, UK: Cambridge University Press.

Craik, K. (1967). *The nature of explanation* (2nd ed.). Cambridge, UK: Cambridge University Press.

da Costa, N. C. A., & French, S. (2003). *Science and partial truth.* Oxford, UK: Oxford University Press.

Driver, R., Guesne, E., & Tiberghien, A. (1985). Children's ideas and the learning of science. In R. Driver, E. Guesne, & A. Tiberghien (Eds.), *Children's ideas in science* (pp. 1–9). Milton Keynes, UK: Open University Press.

Frigg, R. (2010). Models and fiction. *Synthese, 172*(2), 251–268.

Frigg, R., & Hartmann, S. (2009). Models in science. In E. N. Zalta (Ed.), *Stanford encyclopedia of philosophy*. Stanford, CA: Stanford University.

Gentner, D., & Stevens, A. L. (1983). *Mental models*. Hillsdale, NJ: Lawrence Erlbaum.

Giere, R. N. (1988). *Explaining science: A cognitive approach.* Chicago, IL/London, UK: University of Chicago Press.

Giere, R. N. (1999). Using models to represent reality. In L. Magnani, N. J. Nersessian, & P. Thagard (Eds.), *Model-based reasoning in scientific discovery* (pp. 41–57). New York, NY: Kluwer and Plenum.
Giere, R. N. (2002). Models as parts of distributed cognitive systems. In L. Magnani & N. J. Nersessian (Eds.), *Model-based reasoning: Science technology, values* (pp. 227–241). New York, NY: Kluwer and Plenum.
Giere, R. N. (2004). How models are used to represent reality. *Philosophy of Science, 71*, 742–752.
Giere, R. N. (2010). An agent-based conception of models and scientific representation. *Synthese, 172*(2), 269–281.
Gilbert, J. K. (1993). *Models & modelling in science education.* Hatfield, UK: The Association for Science Education.
Gilbert, J. K., & Boulter, C. J. (Eds.). (2000). *Developing models in science education.* Dordrecht, The Netherlands: Kluwer.
Gilbert, J. K., Boulter, C. J., & Rutherford, M. (1998). Models in explanations, Part I: Horses for courses? *International Journal of Science Education, 20*(1), 83–97.
Gilbert, J. K., & Osborne, R. J. (1980). The use of models in science and science teaching. *European Journal of Science Education, 2*(1), 3–13.
Grosslight, L., Unger, C., Jay, E., & Smith, C. L. (1991). Understanding models and their use in science: Conceptions of middle and high school students and experts. *Journal of Research in Science Teaching, 28*(9), 799–822.
Halloun, I. A. (1996). Schematic modeling for meaningful learning of physics. *Journal of Research in Science Teaching, 33*(9), 1019–1041.
Halloun, I. A., & Hestenes, D. (1987). Modeling instruction in mechanics. *American Journal of Physics, 55*(5), 455–462.
Harrison, A. G., & Treagust, D. F. (1996). Secondary students' mental models of atoms and molecules: Implications for teaching chemistry. *Science Education, 80*(5), 509–534.
Hartmann, S. (2008). Modeling in philosophy of science. In M. Frauchiger & W. K. Essler (Eds.), *Representation, evidence, and justification: Themes from Suppes* (pp. 95–122). Frankfurt, Germany: Ontos Verlag.
Hestenes, D. (1987). Toward a modeling theory of physics instruction. *American Journal of Physics, 55*(5), 440–454.
Ingham, A. M., & Gilbert, J. K. (1991). The use of analogue models by students of chemistry at higher education level. *International Journal of Science Education, 13*(2), 193–202.
Johnson-Laird, P. (1980). Mental models in cognitive science. *Cognitive Science, 4*(1), 71–115.
Johnson-Laird, P. (1983). *Mental models.* Cambridge, MA: Harvard University Press.
Justi, R. (2006). La Enseñanza de Ciencias Basada en la Elaboración de Modelos [Modelling-based Science Teaching]. *Enseñanza de las Ciencias, 24*(2), 173–184.
Justi, R., & Gilbert, J. K. (1999). A cause of ahistorical science teaching: Use of hybrid models. *Science Education, 83*(2), 163–177.
Justi, R., & Gilbert, J. K. (2002). Modelling, teachers' views on the nature of modelling, implications for the education of modellers. *International Journal of Science Education, 24*(4), 369–387.
Justi, R., & Gilbert, J. K. (2003). Teachers' views on the nature of models. *International Journal of Science Education, 25*(11), 1369–1386.
Knuuttila, T. (2005a). *Models as epistemic artefacts: Toward a non-representationalist account of scientific representation.* Helsinki, Finland: University of Helsinki.
Knuuttila, T. (2005b). Models, representation, and mediation. *Philosophy of Science, 72*(5), 1260–1271.
Knuuttila, T. (2011). Modelling and representing: An artefactual approach to model-based representation. *Studies in History and Philosophy of Science, 42*(2), 262–272.
Knuuttila, T., & Boon, M. (2011). How do models give us knowledge? The case of Carnot's ideal heat engine. *European Journal for the Philosophy of Science, 1*(3), 309–334.
Kuhn, D. (1991). *The skills of argument.* New York, NY: Cambridge University.

Magnani, L. (2002). Epistemic mediators and model-based discovery in science. In L. Magnani & N. J. Nersessian (Eds.), *Model-based reasoning: Science, technology, values* (pp. 305–329). New York, NY: Kluwer and Plenum.
Morrison, M. (2007). Where have all the theories gone? *Philosophy of Science, 74*(2), 195–228.
Morrison, M. (2011). One phenomenon, many models: Inconsistency and complementarity. *Studies in History and Philosophy of Science, 42*(2), 342–351.
Morrison, M., & Morgan, M. S. (1999). Models as mediating instruments. In M. S. Morgan & M. Morrison (Eds.), *Models as mediators: Perspectives on natural and social science* (pp. 10–37). Cambridge, MA: Cambridge University Press.
Murad, M. H. S. A. (2011). Models, scientific realism, the intelligibility of nature, and their cultural significance. *Studies in History and Philosophy of Science, 42*(2), 253–261.
Nersessian, N. J. (1992a). How do scientists think? Capturing the dynamics of conceptual change in science. In R. N. Giere (Ed.), *Cognitive models of science* (pp. 3–44). Minneapolis, MN: University of Minnesota Press.
Nersessian, N. J. (1992b). In the theoretician's laboratory: Thought experimenting as mental modeling. *Photographic Society of America, 2*, 291–301.
Nersessian, N. J. (1999). Model-based reasoning in conceptual change. In L. Magnani, N. J. Nersessian, & P. Thagard (Eds.), *Model-based reasoning in scientific discovery* (pp. 5–22). New York, NY: Kluwer and Plenum.
Nersessian, N. J. (2002). The cognitive basis of model-based reasoning in science. In P. Carruthers, S. Stich, & M. Siegal (Eds.), *The cognitive basis of science* (pp. 133–153). Cambridge, MA: Cambridge University Press.
Nersessian, N. J. (2008). *Creating scientific concepts*. Cambridge, MA: MIT.
Peschard, I. (2011). Making sense of modelling: Beyond representation. *European Journal for the Philosophy of Science, 1*(3), 335–352.
Pitkin, H. F. (1992). *The concept of representation* (2nd ed.). Berkeley, CA: University of California Press.
Portides, D. P. (2005). Scientific models and the semantic view of scientific theories. *Philosophy of Science, 72*(5), 1287–1298.
Portides, D. P. (2011). Seeking representations of phenomena: Phenomenological models. *Studies in History and Philosophy of Science, 42*(2), 334–341.
Seddon, G. M., & Moore, R. G. (1986). An unexpected effect in the use of models for teaching the visualization of rotation in molecular structures. *European Journal of Science Education, 8*(1), 79–86.
Suárez, M. (2003). Scientific representation: Against similarity and isomorphism. *International Studies in the Philosophy of Science, 17*(3), 225–244.
Suppe, F. (1989). *The semantic conception of theories and scientific realism*. Urbana, IL: University of Illinois Press.
Thagard, P. (2010). How brains make mental models. In L. Magnani, W. Carnieli, & C. Pizzi (Eds.), *Model-based reasoning in science and technology: Abduction, logic, and computational discovery* (pp. 447–461). Berlin, Germany/Heidelberg, Germany: Springer.
Tiberghien, A. (1994). Modeling as a basis for analyzing teaching-learning situations. *Learning and Instruction, 4*(1), 71–87.
van Fraassen, B. (1980). *The scientific image*. Oxford, UK: Oxford University Press.
Vosniadou, S. (2002). Mental models in conceptual development. In L. Magnani & N. J. Nersessian (Eds.), *Model-based reasoning: Science, technology, values* (pp. 353–368). New York, NY: Kluwer and Plenum.
Vosniadou, S., & Brewer, W. F. (1992). Mental models of the earth: A study of conceptual change in childhood. *Cognitive Psychology, 24*(4), 535–585.
Vosniadou, S., & Brewer, W. F. (1994). Mental models of the day/night cycle. *Cognitive Science, 18*(1), 123–183.
Watson, J. D. (1968). *The double helix – The discovery of the structure of DNA*. London, UK: Orion Books.
Weisberg, M. (2007). Who is a modeler? *British Journal for the Philosophy of Science, 58*(2), 207–233.

Chapter 3
Towards Authentic Learning in Science Education

Abstract If a greater emphasis on modelling in the science education curriculum is to be justified, it will have to be as 'authentic' as possible. From the three approaches to authenticity (the child-developmental, the subject-developmental, and the practitioner-situational points of view), a realistic approach must recognise existing subject-content focused practice in science education whilst seeking to attain the maximum engagement by students that is provided by the circumstances of actual scientific practice. The several constraints on such an innovation in schools are identified and existing ways of addressing them are outlined. Particular attention is paid to the identification of contexts that will provide problems capable of being modelled. It is argued that the capability to model authentically will entail a progressive development of experience from school-focused contexts to those that are science- practitioner- focused.

Introduction

In Chap. 1, we argued that, if the current challenges to science education were to be met, the contribution of modelling to their address should be increased. In Chap. 2, we reviewed the philosophical, psychological, and sociological evolution of the notion of 'modelling' itself. Here we take the argument one stage further by considering what 'successful modelling' would entail.

We concluded in Chap. 2 that modelling is, in general, a non-linear process of knowledge construction. It is based on the establishment and integration of three aims: the identification of the purposes which the intended model has to serve; the inclusion of aspects of the modeller's cognitive structure that are derived from personal or impersonal experiences and which are relevant to those purposes; the identification and use of an analogy or mathematical tool that can be used to establish relationships between elements of that experience. The resulting artefact – the model – can be expressed in any one or more of the common modes of representation. These may be: two-dimensional (for example: maps, diagrams, pictures), three dimensional (that is, which are concrete), virtual (that is, which are computer gener-

J.K. Gilbert, R. Justi, *Modelling-based Teaching in Science Education*, Models and Modeling in Science Education 9, DOI 10.1007/978-3-319-29039-3_3

ated), gestural, mathematical, verbal, or a combination of any of these modes. Such a model can serve any one or more of the following epistemic purposes: to represent an entity; to simplify a real-world phenomenon so as to make it amenable to study; to abstract aspects of that phenomenon for closer study; to explain that simplified phenomenon in cause-effect terms; to form the basis for argumentation about the validity of that explanation; to predict its properties under other circumstances; to design future enquiries into the phenomenon; to suggest other models of it.

If we want this view of modelling to have the greatest impact on the challenges that science education faces, we must ensure that the learning that takes place is as *authentic* are possible, that is, that it is as close as possible philosophically, psychologically, and sociologically, to the conduct of science itself. In this chapter, we discuss perceptions of the nature of authenticity in science and science education and identify the limitations to their attainment. Finally, we identify some generic conditions that must be adopted if authenticity is to be attempted: a suitable view of nature of science; a context-based design for the science curriculum; a suitable view of the nature of learning and of its consequences for the processes of teaching.

The Notion of Authenticity in Science Education

Doyle (2000) identifies three meanings for 'authenticity' in education in general. First: *child-centred authenticity*. This involves active physical and mental engagement with

> a task that is inherently meaningful to the child i.e. it is consistent with the child's interests and the child's ways of thinking, knowing, and understanding (p. 2)

The adoption of an entirely child-centred approach to authenticity implies both that:

> the child's conceptions are not seen as misconceptions but rather as real understanding along the developmental path towards maturity (p. 3)

and that

> if adult forms are imposed too early, the child will be forced to engage in meaningless activity without comprehension for external incentives (p. 3).

Complete child-centred authenticity is only met in the early years of primary schooling. It only occurs then, even partially, if teachers' tenaciously hold to their core beliefs that it is in the long-term best interests of the children. This view provides the first justification for the inclusion of modelling in the curriculum, for modelling is a major component of all thinking and hence its exercise is of generic importance to the overall mental development of all children. It has been clearly demonstrated that modelling can be successfully undertaken in child-centred science classes by primary-age (kindergarten) children and that it supports their development (Lehrer & Schauble, 2012).

Second: *subject-centred authenticity*. This

> locates the real in the data, tools, and operations of the disciplines represented as school subjects in the curriculum (p. 1).

This view is the second major justification for the inclusion of modelling in the school science curriculum. It provides a focus on the established ideas of science, which are seen as representing the most important way of understanding the world-as-experienced. Most importantly, the focus is on the ways that such ideas are produced and validated in science. Genuine subject-centred authenticity can be aspired to in the secondary (high) school. However, this aspiration requires persistence, for

> The enemy in this camp is the pre-digested versions of content found in textbooks in which conclusions are given rather than discovered (p. 3).

Third: *situated authenticity*.

> (Here) the real is located in the authentic activity of practitioners in real-world settings'. (…) To learn concepts, students must enter into a community in which the concepts are used and acquire the community's culture by engaging in authentic activity (that is, the ordinary practices of the culture (p. 4–5).

This view of authenticity is perhaps the major justification for the inclusion of epistemic practices related to the production of scientific knowledge, such as those underlying modelling, in the school science curriculum. It is concerned with how the ideas and processes of science are learnt.

For a science education that is truly authentic whilst remaining politically realistic, subject-centred authenticity and situated authenticity must be present simultaneously, they must be intimately interwoven. The subjects of science aspire both to represent and explain the world-as-experienced and to offer access to the better-paid employment that so often requires a knowledge and understanding of some science.

The processes of an authentic science education can be defined as an education in science that is as close as possible to the processes of science itself (Gilbert, 2004). This means an education that:

- supports students' engagement in and learning of both scientific reasoning and the scientific practices involved in the generation and acceptance of scientific knowledge, thus making them understand "“what counts" as a scientific way of generating and validating new ideas" (Windschitl, Thompson, & Braaten, 2008, p. 943);
- recognises and emphasises the role of creativity in the production and use of science in the history of humanity;
- supports the understanding of entities and phenomena in the world-as-experience, that is, contributes to students' views of science as one of the valuable ways of understanding the past, the present, and the future worlds.

This approach to authenticity (summarised in Fig. 3.1) would be manifest in contexts that support 'situated cognition'.

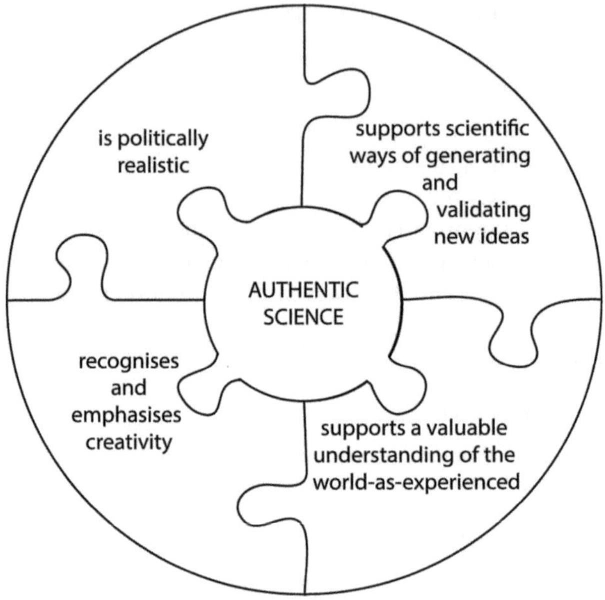

Fig. 3.1 The aspiration to situated cognition

The Nature of Situated Cognition in Science Education

It seems entirely possible that, if an individual were to grow up entirely on their own (admittedly an unlikely situation: *pace* Rousseau's 'Emile'), that person would understand their surrounding environment by use of a self-constructed set of conceptions that were continually in use and under revision in the light of their ability to represent that experience. This type of individualistic cognition is accounted for well by the von Glaserfeld's (1984) theory of 'radical constructivism'. However, almost everybody grows up in a community of some sort. Whilst it is possible to develop entirely individualistic understandings, people normally learn with and from those in their community.

The social constructivist explanation of learning – most closely associated with the work of Vygotsky (1978) – seems to explain how it is that knowledge acquired and retained in a community. The process of knowledge development, which we do believe must be primarily rooted in the mental actions of the individual, is developed and transmitted within a community by means of verbal and non-verbal language expressed in different modes and by the use of technologies, in the course of joint activities. These activities would involve a novice (the student) working in genuine collaboration in a 'cognitive apprenticeship' with an expert (a member of a 'community of practice' built around the activity type in question) (Collins, Brown, & Newman, 1989). This process of mutual knowledge development will only be effective if the participants perceive the activity to be of a joint nature and if the demands of the task lie within the capabilities and interests of the junior partner – within that person's 'zone of proximal development' (Vygotsky, 1978).

These 'communities of practice' can, of course, be centred on any coherent assembly of shared phenomena, activity types, specialised language use, and notions of ethics and standards of performance. In this case, we are concerned with a 'community of scientists' – in practice mainly composed of school science teachers – and the focus is on that collection of phenomena that are of interest to science education, the major concepts that science has produced, the ways in which these were and are developed, together with their implications for technological applications and everyday life. For learning to be legitimately rated as 'situated cognition in science', it must be constructed by the learner co-engaged with an expert in the status of 'legitimate peripheral participation' (Lave & Wenger, 1991), whilst enquiring into phenomena and problems that are, or have been, of genuine interest to scientists.

Roth has provided a concise statement of the requirements that must be met if science activities in school are to be judged as providing the basis for authentic examples of situated cognition:

> School activities, to be authentic, need to share key features with those worlds about which they teach. (…) students should experience scientific inquiry which bears at least five aspects in common with scientists' activities: (1) participants learn in contexts constituted in part by ill-defined problems; (2) participants experience uncertainties, ambiguities, and the social nature of scientific work and knowledge; (3) participants' learning (curriculum) is predicated on, and driven by, their current knowledge state (whatever that might be); (4) participants experience themselves as part of a communities of inquiry in which knowledge, practices, resources, and discourses are shared; and (5) these communities can draw on the expertise of more knowledgeable others, whether they are peers, advisors, or teachers. (Roth, 1995, p. xiii)

As authentic science education, for example that expressed in modelling-based teaching approaches, is not currently widely practiced, we must conclude that all of these conditions are not widely and simultaneously met. Why is this? What can be done to more closely approximate those conditions to the social and psychological conditions that must be met for authenticity to be achieved? Is authentic science education ever possible? It is to these issues that we now turn.

Limitations to the Attainment of Situated Cognition in Science Education

There are a number of reasons why situated cognition is not currently universally practiced in school science classrooms. These reasons may lie to some considerable extent in the philosophical, psychological, and sociological assumptions on which it rests. Abd-El-Khalick (2008) has identified five such assumptions. These, together with the problems in realising each of them in the classroom, are:

- Authentic enquiry is intrinsically motivating and will lead to the sustained engagement of students in learning science.

  Authentic enquiry has not been practiced widely enough in schools for there to be clear evidence of its positive reception by students.

- Children can actually engage in the same kinds of activities undertaken by practicing scientists.

  Whilst young people can display behaviours that are like those of scientists (Gopnik, Meltzoff, & Kuhl, 1999), the social circumstances of schools are different from those of scientific laboratories, for example in respect of relative freedom of action, the use of time, the availability of resources.
- Children can learn in ways that are similar to the ways that scientists conduct authentic enquiries.

  It is possible, indeed likely, that the cognitive processes of children and scientists are similar. They share the common attribute of being human whilst the changes taking place during maturation do not seem too significant. However, the substantial impact of the differing social circumstances between school and the scientific laboratory make an assumption of close similarity in respect of learning difficult to demonstrate (Brewer, 2008). Thus, while scientists are engaging in activities to which they are emotionally committed to a greater or lesser extent, children are effectively just playing a compulsory game devised by the teacher (Jiménez-Aleixandre, Bugallo Rodríguez, & Duschl, 2000).
- The model of scientific enquiry assumed by the science education community is the same as that adopted by scientists.

  Scientific enquiry consists of a large number of components. The existence of a unified model does seem somewhat unlikely given the differences between the distinct sciences (physics, chemistry, biology, amongst others) (Irzik & Nola, 2011). An example of these differences is manifest in the relative possibility of predictive intervention in genetics and in astronomy. In genetics, a scientist can make a change in the genetics of a specimen and see almost immediately if the predicted change in behaviour results. This cannot be done in astronomy. Even if a unified model of scientific enquiry does exist across all the sciences, it does seem unlikely that all scientists will display competence in all of its components. The model of scientific enquiry adopted by the science education community currently does seem to be a creation assembled by 'averaging out' these various factors.
- The science classroom can become a replica of a scientific laboratory.

  The culture of school (for example the use of time, space, the capability for individual action, expectations of performance) is very different from that of a scientific laboratory.

In summarising these concerns, Abd-El-Khalick (2008) concludes that:

> The form of scientific enquiry that simultaneously (a) approximates to scientific enquiry, (b) is viable in pre-college science classrooms, and (c) has the potential to achieve a set of major educational goals, still eludes us. (p. 85)

Truly authentic science education does therefore seem hard to obtain, as represented in Fig. 3.2. However, we can only try to make progress!

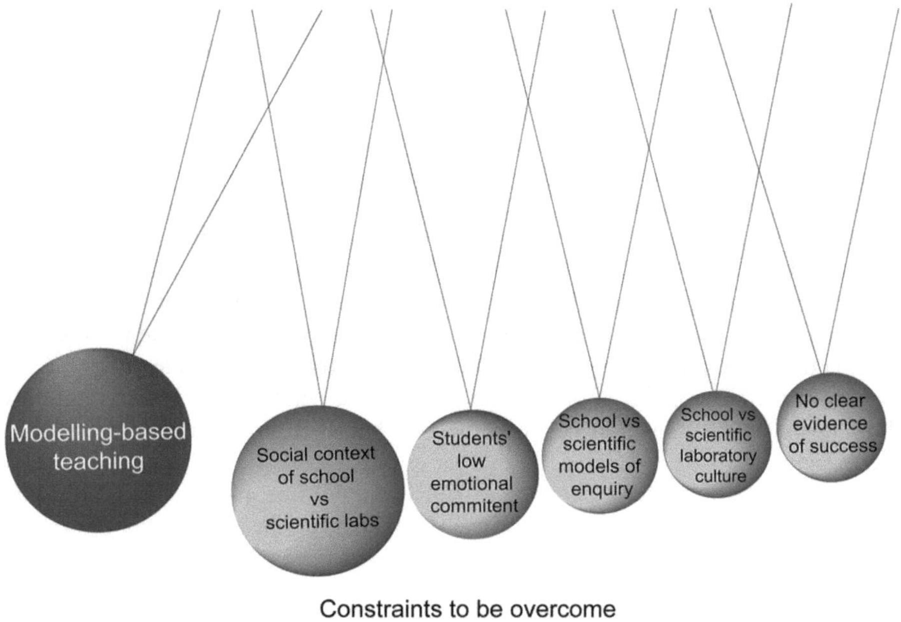

Fig. 3.2 The constraints to be overcome in the pursuit of authenticity

The issue is then how to progress towards the conditions in which authentic scientific enquiry can flourish in schools. Our thesis is that moving towards modelling-based teaching can help do this.

Established Approaches to Facilitate Modelling

A realistic way of bringing about educational change is to identify and retain those elements in current practice that are congruent with the desired innovation whilst introducing necessary changes in those which are not. To a first approximation, this will involve identifying and categorising all the elements of existing pedagogic practice. Bernstein (1990) identified two basic forms of pedagogy in schools. On the one hand, there is the 'visible' pedagogy: the explicitly stated, tangible, readily acknowledged, transformations of knowledge and behaviour from the everyday or academic world into the forms in which they are manifest only in schools. For example, in respect of: the roles of the participants (teachers, students) in what is done; the content of what is being learnt; the notions of progress in that learning. On the other hand, there is the 'invisible' pedagogy: those transformations that exist but are not recognised or acknowledged in any explicit way. For example in respect of: the 'factory model' of classroom organisation, the structure of the 'working day', and the distribution of talk between the teacher and the students. The attainment of authenticity implies that those elements contained within the invisible pedagogy are made visible and, in most cases, substantially changed. This is not an easy task: the evidence is that has, so far, only been partially achieved and in favourable circumstances.

Although it is to some extent a parody, long-established, customary practice in science classrooms seem to assume that all existing scientific knowledge is true and unchanging. This assumption is an act of belief, such that the provision of detailed evidence in support of its acceptance is not thought necessary, and such that it remains part of the 'invisible' pedagogy. The result of this belief is that the ideas of science are customarily taught as a 'rhetoric of conclusions', that is, as if their acceptance requires no evidence of justification. The assumptions (and their associated beliefs!) of situated cognition differ, very greatly, from these assumptions, as we have seen. The issue, in the pursuit of authentic science education, is not only how to make the transition between the two but also how to make that change permanent.

Attempts to do so have been made. For example, Lee and Songer (2003) collected together four discrete ideas for ways of increasing the authenticity of existing practical work in science. In the case of the four, we have added a comment on their implications for change:

- Students could address real-world problems of the type tackled by scientists.
  This would entail the nature of such problems being appreciated by school science teachers.
- Students could seek solutions to problems encountered in their everyday lives.
  This would entail students becoming aware of those problems in everyday life to which scientific ideas and methodology had some relevance.
- Linking students and scientists through data sharing and the mutual critiquing of work, both at a distance and through direct communication.
  This would entail the notion of the 'school as refuge, fortress, or prison' being abandoned.
- Adding elements of scientists' authentic activities to students' work.
  For example: engaging in argumentation, presenting their work to other students, organising communication between groups of students and scientists.

The implied changes are considerable.

On the other hand, Lee and Songer also opine that aspects of authentic enquiry can actually be transformed until they can be carried out in schools. This would involve:

- Lowering content-knowledge demands.
  This might be done by using lower-level (historical) causal models of phenomena where these provide routes to adequate explanations. For example, 'heat transfer' models of energy seem to be accessible to students and hence make learning in the lower high-school grades easier than that expected by the use of 'molecular-kinetic' models.

- Lowering demands for scientific process skills.
  This could involve the use of simplified models to guide the enquiry process. For example, in some contexts, the sequence: 'question, predict, experiment, model, apply' could be so used.
- Lowering the resource demands of authentic enquiry.
  Computer-based sources could be used for carrying out aspects of an enquiry, for example in order to produce visualisations of data by the application of modelling-software.

Chinn and Malhotra (2001) also took a systematic approach to a demand-reduction approach, arguing that, because schools were unable to provide the time, resources, or expertise, within which to facilitate fully authentic scientific enquiry, simpler versions of such activities should be developed. They then developed lists of both the cognitive and the epistemological processes involved in authentic scientific enquiry. Whilst the word 'modelling' does not appear in their list of dimensions of epistemology, the word 'model' occurs twice. Under *purpose of research*, authentic enquiry is said to involve tasks in which "Scientists aim to build and revise theoretical models with unobservable mechanisms" (p. 188). Under *theory-data coordination*, authentic tasks involve those in which "Scientists coordinate theoretical models with multiple sets of complex, partially conflicting data" (p. 188). The list of cognitive and epistemological processes – including the two that mention models – were then used to evaluate both standard existing textbook practical tasks, such as:

> *simple experiments* (in which) students conduct a straightforward experiment, usually evaluating the effects of a single independent variable (…) *'simple observations'* (where) students carefully observe and describe objects. (…) *'simple illustrations'* (where) students follow a specified procedure, usually without a control condition, and observe the outcome (Chinn & Malhotra, 2001, p. 179, original emphases).

Some tasks were also especially developed by educational experts with the idea of specifically supporting the provision of authentic scientific enquiry in the classroom. The conclusions that must be drawn from this analysis and attempts at synthesis are that shaping tasks such that they facilitate authentic scientific enquiry in the classroom is a very complex and demanding task.

The fact that these – and other – published accounts of minor and partial attempts to move towards 'authentic science education' show little evidence of success suggests that many elements of the curriculum must be simultaneously changed. In some cases, using the language of Bernstein (1990), all the elements of the curriculum must be made 'visible', with some of them being modified or replaced. For the greatest chance of success, a number of major elements must certainly be addressed. These concern: the provision of suitable contexts for study with in-built ill-defined problems; enabling students to experience the social nature of scientific work by participating in 'communities of scientific practice; basing the practice of modelling on students' existing knowledge and skills. We now take each of these elements in turn. In each case, the issues are: what is the desired pedagogic structure? If we assume a general model of existing practice that substantially differs from the desired structure, what changes have to be made to it?

The Provision of Suitable Contexts to Study

The word 'context' is often used loosely when discussing educational matters, sometimes being conflated in meaning with 'situation' or 'general example'. Duranti and Goodwin (1992) saw a context as having four attributes. First, it is formed around a 'focal event' – some important or typical entity or event. Thus the existence and operation of an oil refinery in a particular place is an entity, as is the holding of the Olympic Games. Second, that focal event must be available in detail in some way, for either direct or indirect experience, so that those engaged in an educational activity can discuss it amongst themselves. Thus a class might visit to an oil refinery, or discuss a video of the Olympics. Third, the focus entity chosen must be such that the specialist concepts entailed in its analysis must be readily invoked in that discussion. Where a wide range of potential contexts can be identified, meeting this criterion is of the greatest importance. Fourth, the discussion of the focal entity must enable those broader aspects of the language register of science to be readily introduced. This discussion will not only provide a forum for the introduction of new terms but also enable existing misconceptions to be remedied. We would add a fifth attribute to this list: any context chosen must be such students, having studied it, can readily transfer their newly-acquired conceptual understanding to other contexts (Schwartz, Varma, & Martin, 2008).

Existing practice in respect of the notion of 'context' often deviates quite considerably both from that which would meet the five attributes and do so in such a way as to facilitate 'authentic science education'. First, while much school science is actually taught within the framework of focal events, the problem is that such focal events are too often those that were contained in the original reports of the research that established the concepts at their heart. To take two examples: looking at the behaviour of light within the framework of the 'optical bench'; looking at electromagnetic phenomena using the coiled wire/magnet equipment of Faraday. These traditional 'contexts' do not map on to the everyday experience of students. Second, although electronic media have made surrogate access to a wide variety of contexts possible, schools still persist in using those to which they have immediate, concrete, access. Third, many of the educationally most important concepts, for example in nanoscience, can only be readily evoked within modern contexts: these are not often readily accessible. Fourth, the lack of appropriate contexts inevitably hinders the deployment of specialist language. Fifth, the traditional contexts that are commonly used show little evidence of the transfer of concepts to those contexts that are of importance in the modern world. Good practice in overcoming all of these shortcomings is no doubt widely established but is poorly documented.

Students Experiencing the Social Nature of Scientific Work in 'Communities of Scientific Practice'

All knowledge that has been found, for one reason or another, to be so worthwhile as to be communicated in writing to succeeding generations, is grouped into 'disciplines'. In their seminal work on the nature of 'disciplines of knowledge', King and Brownell (1966) described a discipline as having eight characteristics. It:

- exists in communities of people with a sustained interest in it;
- uses imagination in a particular, characteristic, way;
- is exercised on a particular domain of phenomena of interest;
- has a tradition of what constitutes 'good practice' in that domain;
- uses a particular mode of enquiry, that is, it has a syntactical structure;
- builds up a corps of concepts, models and theories, that is, it has a substantive structure;
- develops a specialist language with which the meaning of its substantive structure can be expressed;
- accumulates a framework of beliefs about what might and should be known within it and about the world-as-experienced.

The notion of a 'community of scientific practice' is at the heart of these characteristics, even though many themes in modern science involve several disciplines, for example genetics involves both chemistry and biology. For students to be successfully inducted into any discipline or disciplines – to be able to engage in authentic activity – they must work collaboratively with established practitioners – experts – in the discipline. By so-doing, they will acquire a sense of the scope and boundaries of the phenomena addressed by the discipline(s) in question, expertise in the ways in which enquiry may properly be carried out, an understanding of its specialist language and hence an understanding of its key ideas, and, finally, an appreciation of the beliefs that encapsulate the legitimate scope and value of the whole.

At the core of any 'community of scientific practice' is the notion of practical work and all its attendant mental activities. Practical work, as carried out in most schools at present, for example as reviewed by Hodson (1990), only presents a very limited range of opportunity to experience and acquire scientific expertise. The reasons why this is so stem from the organisational underpinnings of school science. Thus, to partially revisit and extend some of the ideas presented earlier in this chapter:

- in general, science teachers have not themselves conducted scientific enquiries, and thus cannot be said to be experts in science per se;
- the laboratory resources available to schools are far less sophisticated than those available to practicing scientists;
- in schools, students generally work either on their own or in small groups which are often not to some considerable extent of their choosing;
- the phenomena on which students work are neither chosen by them nor are they usually allowed to continue to work on any phenomenon for a reasonable time;

- the opportunities for acquiring and deploying specialist language are very limited.

If school science is become more authentic, this entails:

- either science teachers becoming more expert in the practice of science or students being brought into sustained contact with scientists;
- either students having access to better equipped laboratories or the contexts about which they study must be amenable to extensive enquiry with existing resources;
- students being able, indeed expected, to work in groups for a sustained period of time;
- developing greater opportunities for acquiring and deploying specialist language.

The demands attached to providing a fully 'community of practice' are indeed great: Ben-Ari (2005) believes that they are impossible to meet, not least because of the narrowing effect on students' learning that would enable them to relate in depth to that of the expert to whom they were 'apprenticed'. However, efforts – evidently successful to some degree – have been made to approximate the learning circumstances to those of a 'community of practice'. Barab and Hay (2001) have outlined the extremes of what are possible: the 'simulated environment', in which the school classroom is adapted to more closely resemble the environment of a research laboratory; and the 'participation environment', in which partial but genuine engagement in research is possible.

Moving away from the fully simulated environment towards a more participatory environment, Helms (1998) describes a project conducted by a class of 9th graders on a wetland area: the work was managed in detail by the class science teacher, but a scientist was fully engaged in the work. Continuing the trajectory of that movement towards a more participatory environment, in the case study reported by Barab and Hay (2001), middle school students, their teachers, and scientists, participated in a 2 week long 'camp' (one must assume that it was non-residential) in which a series of fairly open-ended projects on biological phenomena were undertaken. Moving sharply towards meaningful community participation, Bouillion and Gomez (2001) describe a project in which teachers of several school subjects, 5th grade students, joined with parents, others from the community, and scientists, to investigate and make recommendations about an environmental problem in their district. Robertson (2007) reports a somewhat similar activity, but here the emphasis was placed on the conditions under which the collaboration would be most successful. These are that the participants:

- Dedicate ample time towards collaboration in order to effectively communicate and get to know each other;
- Communicate openly and often in order to gain understanding, as well as build trust and relationships;
- Strive to understand each other's perspective so that they can be incorporated into the shared vision;

- Find ways to foster the motivation and ownership that is necessary for persevering through negotiations towards a shared vision;
- Cultivate a positive and encouraging collaborative environment. (Robertson, 2007, p. 701).

All these cases involved younger students and focused on activities in which relatively little detailed knowledge of the ideas of science was needed. Perhaps inevitably, given the level of detailed scientific knowledge needed to enter meaningfully into a 'community of practice', the few case studies reported involve extensive interaction between practicing scientists, older students and biological phenomena. Charney, Hmelo-Silver, Sofer, Neigeborn and Nemeroff (2007) discuss an extended project in which 10th and 12th grade students were engaged in a project built around a carefully designed phenomenon involving molecular genetics. In the most 'advanced' case study to come to our attention, Hsu (2010) reports a project for 11th grade biology students that involved them and their science teachers in 'internship' in a university biology department. The overall conclusion to be drawn from these case studies is that approximations to genuine 'communities of practice' are perhaps only realistic, in the absence of other examples, for older students and 'biological' phenomena. However, the process of introducing all students to the cultural contexts of science can and should begin at a much earlier age.

Basing Modelling on Students' Existing Knowledge and Skills

In addition to using 'contexts' as the basis for science education and to structuring learning as closely as possible on that taking place in 'communities of practice', teachers need to know several things about their students. They must be conversant with: how students' reason, for scientific enquiry is based on the use of special kinds of reasoning, like the use of logic and analogy; what students believe a model to be, for this will govern the nature of their intellectual product, if authentic modelling activities are to be provided. Teachers may not be fully, or even partially, satisfied that the answers they receive are those that will permit the development of the skills of modelling. However, the sustained and thorough implementation of modelling-based teaching (as discussed in Chap. 4), should bring students' knowledge and skills progressively closer to what are required for fluent and accurate modelling: the notion of 'learning progression' (that is the subject of Chap. 10).

Although the legitimacy of the analogy that students, even young children, 'think like scientists' has its limitations (Brewer, 2008) and is still much contested (for instance, by Gopnik, 1996), it has proved a positive influence on science education (Brewer, 2008). So teachers may, with reasonable confidence, assume that their students will be able to reason in ways that enable them to engage in relevant epistemological practices, like modelling. Where doubts arise in practice, for example concerning the relation between a 'cause' and an 'effect', the teacher will have to take appropriate remedial action on the spot.

As the ultimate objective of a modelling activity is to produce a model, it is therefore crucial that students have a clear understanding of what a model is. The earliest work on students' understanding of the nature of 'model', undertaken on 12–13 year olds and 16–17 year olds and some practicing scientists by Grosslight, Unger, Jay and Smith (1991), found three levels of understanding: Level 1, where a model is regarded as an incomplete copy of reality; Level 2, where a model is produced for the purposes of communication, some aspects of reality have been omitted, diminished, or enhanced; Level 3, where a model is perceived to be an intellectual creation, produced to help the development of scientific explanations. None of the students had reached Level 3. However, Svoboda and Passmore (2013) have observed that:

> these results (Grosslight et al., 1991) represent students' unscaffolded, decontextualized, ideas about models and do not reflect what students can do with models in more supportive or appropriately contextualised settings. Our interpretation of Grosslight et al.'s findings is that the student responses they collected reflect the simplistic definitions of models that are typically presented in school, but are not representative of what students can do when engrossed in more authentic modelling activities (p. 136–137).

Whilst developmental work with school-age students has shown encouraging, for example in the work of Keys (1995) and of Raghavan and Glaser (1995), a study of student teachers showed that, even after the use of a specially designed module on the notion of 'model', they only progressed to Level 2 (Crawford & Cullin, 2004). When added to the work of Justi and Gilbert (2003) on practicing teachers in Brazil, the work on teachers suggests that teacher development in respect of models and modelling is essential if authentic science education is to be widely provided. This theme is thus the focus of Chap. 11.

From the late 1970s onwards, there was a huge output of studies into the meanings that students (and many of their teachers!) had for the words used in science, for example 'force', 'energy', and which were derived from everyday experience and folklore rather than from science education. Pfundt and Duit (2000) have produced a bibliography listing many hundreds of such studies. These meanings were given several titles, of which one was 'alternative conceptions' (Gilbert & Watts, 1983). Considerable international effort was made to find ways in which these alternative conceptions could be altered to or replaced by the scientifically accepted meaning. That by Posner, Strike, Hewson and Gertzog (1982) has proved very influential, although it has been criticised as being too logical and of paying too little attention to the affective, motivational, contextual aspects of conceptual change (Pintrich, Marx, & Boyle, 1993). Detailed consideration of the nature of these alternative conceptions and of their amendment or replacement is beyond the scope of this volume. However, it does seem that modelling activities will bring about desired changed meanings for constituent concepts (see Chap. 10).

In this chapter we have discussed the nature of authentic science education and pointed to the indicators and social circumstances that lead to a first approximation of its provision. In the next chapter we present some approaches to the implementation of modelling-based teaching in the pursuit of authentic science education.

References

Abd-El-Khalick, F. (2008). Modeling science classrooms after scientific laboratories. In R. Duschl & R. E. Grandy (Eds.), *Teaching scientific inquiry: Recommendations for research and implementation* (pp. 80–85). Rotterdam, The Netherlands: Sense.

Barab, S. A., & Hay, K. E. (2001). Doing science at the elbows of experts: Issues related to the science apprenticeship camp. *Journal of Research in Science Teaching, 38*(1), 70–102.

Ben-Ari, M. (2005). Situated learning in 'This high technology world'. *Science & Education, 14*(3–5), 367–376.

Bernstein, B. (1990). *The structure of pedagogic discourse: Class, codes and control*. London, UK: Routledge.

Bouillion, L. M., & Gomez, L. M. (2001). Connecting school and community with science learning: Real world problems and school-community partnerships as contextual scaffolds. *Journal of Research in Science Teaching, 38*(8), 878–898.

Brewer, W. F. (2008). In what sense can the child be considered to be a "little scientist"? In R. Duschl & R. E. Grandy (Eds.), *Teaching scientific inquiry: Recommendations for research and implementation* (pp. 38–49). Rotterdam, The Netherlands: Sense.

Charney, J., Hmelo-Silver, C. E., Sofer, W., Neigeborn, S., & Nemeroff, M. (2007). Cognitive apprenticeship in science through immersion in laboratory practices. *International Journal of Science Education, 29*(2), 195–213.

Chinn, C. A., & Malhotra, B. A. (2001). Epistemologically authentic inquiry in schools: A theoretical framework for evaluating inquiry tasks. *Science Education, 86*(2), 175–218.

Collins, A., Brown, J., & Newman, S. (1989). Cognitive apprenticeship: Teaching the craft of reading, writing and mathematics. In L. Resnick (Ed.), *Cognition and instruction: Issues and agendas* (pp. 453–494). Hillsdale, NJ: Lawrence Erlbaum.

Crawford, B. A., & Cullin, M. J. (2004). Supporting prospective teachers' conceptions of modeling in science. *International Journal of Science Education, 26*(11), 1379–1401.

Doyle, W. (2000). *Authenticity*. Paper presented at the American Educational Research Association Conference, Montreal, QC, Canada.

Duranti, A., & Goodwin, C. (Eds.). (1992). *Rethinking context: Language as an interactive phenomenon*. Cambridge, UK: Cambridge University Press.

Gilbert, J. K. (2004). Models and modelling: Routes to a more authentic science education. *International Journal of Science and Mathematics Education, 2*, 115–130.

Gilbert, J. K., & Watts, D. M. (1983). Conceptions, misconceptions, and alternative conceptions: Changing perspectives in science education. *Studies in Science Education, 10*(1), 61–98.

Gopnik, A. (1996). The child as scientist. *Philosophy of Science, 63*(4), 485–514.

Gopnik, A., Meltzoff, A., & Kuhl, P. (1999). *The scientist in the crib: Minds, brains, and how children learn*. New York, NY: William Morrow.

Grosslight, L., Unger, C., Jay, E., & Smith, C. L. (1991). Understanding models and their use in science: Conceptions of middle and high school students and experts. *Journal of Research in Science Teaching, 28*(9), 799–822.

Helms, J. V. (1998). Science and/in community: Contexts and goals in practical work. *International Journal of Science Education, 20*(6), 643–653.

Hodson, D. (1990). A critical look at practical work in school science. *School Science Review, 71*(256), 33–40.

Hsu, P.-L. (2010). From a sense of stereotypically foreign to belonging in a science community: Ways of experimental description about high school students' science internship. *Research in Science Education, 40*(3), 291–312.

Irzik, G., & Nola, R. (2011). A family resemblance approach to the nature of science for science education. *Science & Education, 20*(7–8), 591–607.

Jiménez-Aleixandre, M. P., Bugallo Rodríguez, A., & Duschl, R. (2000). "Doing the lesson" or "doing science": Argument in high school genetics. *Science & Education, 84*(6), 757–792.

Justi, R., & Gilbert, J. K. (2003). Teachers' views on the nature of models. *International Journal of Science Education, 25*(11), 1369–1386.
Keys, C. W. (1995). An interpretative study of students' use of scientific reasoning during a collaborative report writing intervention in ninth grade general science. *Science Education, 79*(4), 415–435.
King, A., & Brownell, J. (1966). *The curriculum and the disciplines of knowledge: A theory of curriculum practice*. New York, NY: Wiley.
Lave, J., & Wenger, E. (1991). *Situated learning: Legitimate peripheral participation*. Cambridge, UK: Cambridge University Press.
Lee, H.-S., & Songer, N. B. (2003). Making authentic science accessible to students. *International Journal of Science Education, 25*(8), 923–948.
Lehrer, R., & Schauble, L. (2012). Seeding evolutionary thinking by engaging children in modeling its foundations. *Science Education, 96*(4), 701–724.
Pfundt, H., & Duit, R. (2000). *Bibliography: Students' alternative frameworks and science education* (5th ed.). Kiel, Germany: Institute of Science Education, University of Kiel.
Pintrich, P. R., Marx, R. W., & Boyle, R. A. (1993). Beyond cold conceptual change: The role of motivational beliefs and classroom contextual factors in the process of conceptual change. *Review of Educational Research, 63*(2), 167–199.
Posner, G. J., Strike, K. A., Hewson, P. W., & Gertzog, W. A. (1982). Accommodation of a scientific conception: Toward a theory of conceptual change. *Science Education, 66*, 211–227.
Raghavan, K., & Glaser, R. (1995). Model-based analysis and reasoning in science: The MARS curriculum. *Science Education, 79*(1), 37–61.
Robertson, A. (2007). Development of a shared vision: Lessons from a science education community collaboration. *Journal of Research in Science Teaching, 44*(5), 681–705.
Roth, W.-M. (1995). *Authentic science education: Knowing and learning in open-inquiry science laboratories*. Dordrecht, The Netherlands: Kluwer.
Schwartz, D. L., Varma, S., & Martin, L. (2008). Dynamic transfer and innovation. In S. Vosniadou (Ed.), *International handbook of research on conceptual change* (pp. 479–506). New York, NY: Routledge.
Svoboda, J., & Passmore, C. M. (2013). The strategies of modeling in biology education. *Science & Education, 22*(1), 119–142.
von Glaserfeld, E. (1984). An introduction to radical constructivism. In P. Watzlawick (Ed.), *The invented reality* (pp. 17–40). New York, NY: Norton.
Vygotsky, L. S. (1978). *Mind in society: The development of higher psychological processes*. Cambridge, MA: Harvard University Press.
Windschitl, M., Thompson, J., & Braaten, M. (2008). Beyond the scientific method: Model-based inquiry as a new paradigm of preference for school science investigations. *Science Education, 92*(5), 941–967.

Chapter 4
Approaches to Modelling-Based Teaching

Abstract To clarify the language used in the field, the distinction is drawn between 'model-based teaching' (the use of existing models by students) and 'modelling-based teaching' (MBT) (the creation and use of models by students). The range of activities that can be included in the two forms is reviewed and five generic types briefly described. The two types concerned with MBT are then discussed at some length and existing approaches to 'learning to model *de novo*' are reviewed. The use of the 'Model of Modelling' (MM) to base science teaching is discussed at some length and the skills that may be developed at each of its stages are established. The design and implementation of teaching sequences using MBT is presented with the use of case-study material. Finally, evidence of the impact on student learning of MBT conducted through the use of the MM is presented.

Authentic science education has been defined as an education in science that is as close as possible to science itself (Gilbert, 2004). This means an education that:

- supports students' engagement in and learning of both scientific reasoning and the scientific practices involved in the generation and acceptance of scientific knowledge, thus making them understand "what counts as a scientific way of generating and validating new ideas" (Windschitl, Thompson, & Braaten, 2008, p. 943);
- recognises and emphasises the role of creativity in the production and use of science in the history of humanity;
- supports the understanding of entities and phenomena in the world-as-experience, that is, contributes to students' views of science as one of the valuable ways of understanding the past, the present, and the future worlds.

As modelling is one of the essential processes in producing, validating, disseminating, and using scientific knowledge, science education researchers have emphasised the importance of its inclusion in science education contexts (for instance, Acher, Arcà, & Sanmartí, 2007; Barab, Hay, Barnett, & Keating, 2000; Campbell, Oh, & Neilson, 2012; Clement, 1989; Gilbert, 2004; Gobert & Buckley, 2000; Halloun, 2007; Justi & Gilbert, 2002b; Lehrer & Schauble, 2012; Schwarz et al., 2009; Svoboda & Passmore, 2013; van Joolingen, 2004). Additionally, the recent 'Framework for K-12 Science Education' (National Research National Research Council, 2012) places the design and use of models as one of the eight 'scientific

J.K. Gilbert, R. Justi, *Modelling-based Teaching in Science Education*, Models and Modeling in Science Education 9, DOI 10.1007/978-3-319-29039-3_4

and engineering practices used to establish, extend, and refine knowledge' – one of the three major dimensions around which science education is expected to be built.[1]

The use of modelling activities in teaching characterises what has been named *modelling-based teaching* (MBT). In a broader sense, it is defined as

> any implementation that brings together information resources, learning activities, and instructional strategies intended to facilitate mental model-building both in individuals and among groups of learners (Gobert & Buckley, 2000, p. 892).

MBT aims at contributing to the active engagement of students in their learning process as it enables them to create models, to express them, to assess both their own models, those produced by their colleagues and/or introduced by the teacher, and by arguing about all of them. As it is discussed in this chapter, MBT characterises socio-interactive constructivist teaching contexts in which knowledge building results from student(s)-student(s), student(s)-teacher, and student-resource interactions. This means that MBT may support co-construction of knowledge. One of the main outcomes of this process is that students may not only build their knowledge from the establishment of relationships between multiple ideas (obtained from multiple sources) – *learning science* in Hodson's (1992, 2014a) terms – but also understand modelling and its role in the development of scientific knowledge. Thus, the involvement of students in MBT helps to promote an understanding that goes beyond the rote memorization of facts and information. This also tends to help the development of a flexible and critical knowledge about scientific concepts, students' meta-knowledge about modelling (also called meta-modelling), a series of higher order thinking skills that can be used in different situations and problems (including the ability to ask questions), as well as the understanding of specific aspects regarding the nature of science – elements of what Hodson (1992, 2014a) named *learning about science*. In one of his most recent publications, Hodson (2014a) also emphasises two other core goals for science education: *doing science*, for which MBT contributes by providing opportunities for students of participating in, and reflecting on, the practice of modelling; and *addressing socio-scientific issues*, which may occur when authentic modelling practices contexts (those that involve real-world or socio-scientific problems) are used as MBT contexts (Prins, Bulte, van Driel, & Pilot, 2008).

In this chapter, the aspects involved in MBT are discussed. The main possible outcomes from the use of MBT, although generally discussed in this chapter, are detailed in other chapters. The main arguments are supported by results from previous research in which MBT occurred in different ways.

[1] The others being crosscutting concepts (those that have application across all domains of science), and disciplinary core ideas (those that can support students future acquisition of information on their own) (National Research National Research Council, 2012).

Relevant Distinctions

Before proceeding, it is necessary to draw some important and useful distinctions. Initially, in the literature, the acronym MBT is used as meaning both "model-based teaching" and "modelling-based teaching" (these are sometimes used simultaneously). Additionally, some authors use "model-based teaching" as synonymous with "modelling-based teaching", that is, meaning a process in which students participate in the practice of modelling. In order to avoid misunderstandings, we assume *model-based teaching* to be an instructional process in which models provided by the teacher, the textbook or any other source are *used by students*, and *modelling-based teaching* as an educational process in which *students are involved in modelling a given entity*. However, it is important to remember that, according to constructivism, we never learn things by rote. Rather, we only succeed or fail in constructing our own understandings of what we are taught. So we acknowledge that one does 'construct a mental model' in *model-based teaching*, but the level of creativity and complexity of the required cognitive processes in that case is far lower than that involved when learning during *modelling-based teaching* (as clearly emphasised by Morgan (1999) in the context of economics). Being consistent with the core focus of this book, *modelling-based teaching* is the process mainly discussed in this chapter, independently of how authors named it.

Second, some authors propose the existence of distinct approaches for MBT (that sometimes are identified as 'types of learning activities in modelling' (Mellar, Bliss, Bliss, & Boohan, 1994), 'types of learning in reference to modelling' (Tiberghien, 1994), 'types of modelling' (van Joolingen, 2004), or other similar phrases. In the following paragraphs, we present a summary of the main original proposals, following by a synthesis that guides the discussion in the rest of the chapter.

In one of the first books focused on computer-based modelling, Mellar et al. (1994) identified two types of learning activities in modelling:

- *exploratory* ones, in which students work with someone else's models (for instance, models presented by teachers or textbooks) that are – at least initially – often quite different from their own models; and
- *expressive* ones, in which students create their own models, thus externalising their own ideas and supporting their reflection on them.

More recently, when discussing the roles of modelling in inquiry learning, van Joolingen (2004) expressed his view that exploratory and expressive modelling are two ends of a continuum that involves the use of variables and their relations in models. From this perspective, exploratory modelling is characterised as a situation in which both the variables and relations are fixed (since they were defined by the author of the model) and the students' main aims are to discover the properties of a model and/or to investigate the structure of a model, whilst in expressive modelling variables and relations are defined by the modeller, who intends to express his/her own ideas. van Joolingen also identifies another point in that continuum which he termed *inquiry modelling*, a situation in which the main aim is to reconstruct a model (which implies that most variables are those from the target model, and the rules are defined by the modeller).

In the context of physics, Tiberghien (1994) assumes that interpretation and prediction imply a modelling process operating at three levels: theory (concerning explanations, that is, constituted by paradigms, principles, laws, etc.), model (consisting of qualitative and quantitative relations associated with observable phenomena and physical quantities aiming at representing selected aspects of the phenomena), and experimental field of reference (involving the experimental situations concerning the validity of the theoretical constructions, that is, involving experimental facts, experimental devices, measurements). Thus, modelling involves the establishment of functional relations between theory and the material world through the role of models. From such ideas, she then discusses how students learn physics by constructing models that allow them to interpret and/or make predictions about material situations, and propose four types of learning:

- *extension of the field of applicability*, when new elements are added only at the level of events, that is, there is no change in the students' knowledge at the level of theory and model. For instance, students know that when water is heated, it becomes hot, boils, and evaporates. After teaching, they only include the idea that there is a maximum temperature involved in the phenomena. The linear causality between the source of heating and the objects/material that is heated is not changed;
- *new semantic relations between model and the field of applicability*, when students use a recently learned model in a context different from the one used to teach it, which requires a new organisation of their knowledge. For instance, after learning that heat flow depends on the substance, a student was able to interpret heating and insulating by emphasising the role of the agent that causes the process rather than by focusing on the object in which the effect is observed, as she had done previously;
- *new model in relation to a new theoretical level*, that occurs when, as a result of a given model be applied in a different context, some elements of both the model and the theory that supports it are restructured in a way that is coherent with physics. For instance, when trying to explain the insulation of an object, a student was able to differentiate two physical quantities – heat and temperature – instead of one. In other words, the student constructed an acceptable meaning for heat and temperature, that is, a theoretical conceptual change occurred;
- *theory constructed from social teaching rules*, which occurs when students are able to use a model in a different context, but without knowing the theory that supports it. In fact, they develop a theory based on the rules of using the model. According to Tiberghien, a typical example is that of a student who "correctly solves all the calculus but are not able to use their results to predict real events (relevant for the calculus)" (Tiberghien, 1994, p. 85). This means that the student do not really understand the physics phenomena.

In our own previous work (Gilbert, 2004; Justi & Gilbert, 2002a), we identified five approaches to learning about models and modelling in the literature. They were characterised as:

- *learning curricular models*, a situation in which the focus is on the content of science, that is, in which students have to learn a product – the curricular model – sometimes with the help of teaching models (devices developed specially to reach such a goal). This occurs, for instance, when a student learn about the structure of the atom as proposed by Niels Bohr by making analogical relationships between the solar system and such a model;
- *learning to use models*, when, after being presented to a given model, students have the opportunity to apply it in a context where it will works, for instance, by being successful in adequately explaining a phenomenon. For instance, when a student learns how a solid compound dissolves in water from the perspective of the particulate nature of matter's model, s/he is then asked to explain how, for example, the air is filled with the smell of a flower;
- *learning to revise models*, when students have the opportunity to change a model so that it can be used in a context distinct from the one initially encountered and/or it can be used for purposes distinct from its original ones. For instance, having learnt about a model for controlling a disease outbreak in a given environment, students can revise it to explain possible differences in the process when they occurred in distinct environments;
- *learning to reconstruct a model*, when students recreate a model whose essence they already known by using modes of representation other than the one originally used to express the model, and in such a creative way that some unknown details or relationships involving the model are also learnt. This is very common in physics teaching, when students are asked to produce mathematical models related to qualitative models that they already know for explaining a given phenomenon;
- *learning to construct a model de novo*, a situation in which students work without knowing the outcome model beforehand. This means that students have to think about the entity being modelled by asking and answering questions about it, deciding on the best way to build the model, planning and conducting tests of the model, and evaluating the overall outcome. Therefore, this is a situation that may occur when students are learning about the subject of the model for the first time, and the teacher provides proper conditions and support for the engagement of students in the epistemic practices involved in modelling.

The characterisation of each of these approaches does not mean that they are mutually exclusive. Modelling as a whole process may include movements backwards and forwards across some of them in both scientific and educational contexts, as synthesised in Fig. 4.1. In some sense, it seems "natural" to involve students in simpler approaches before they can be able to perform the whole process of constructing models.

The proposals previously summarised have some points of intersection. The clearest ones are:

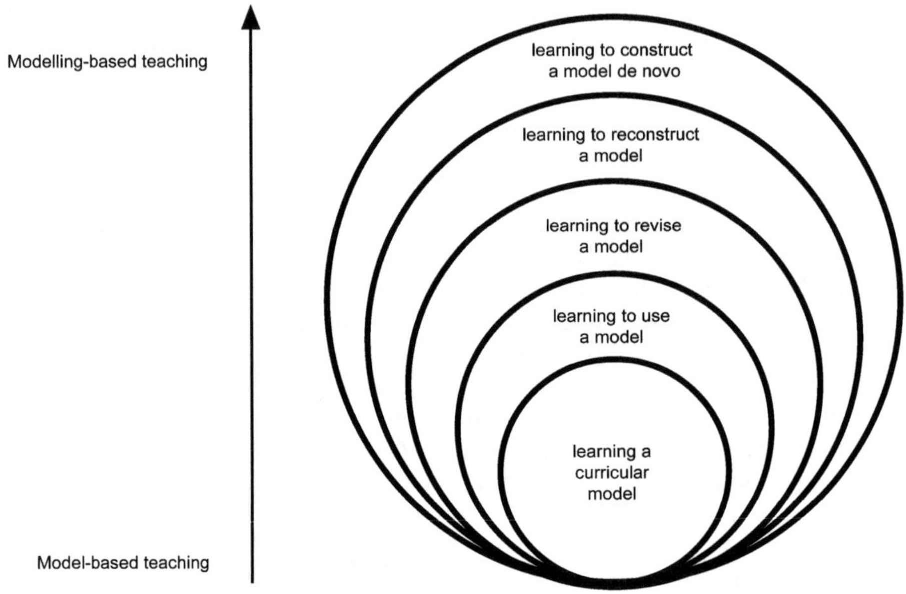

Fig. 4.1 Approaches to learning about models and modelling

- *Learning to use models* and *exploratory learning* have the same meaning. In particular, they emphasise that students work with models, but are not involved in modelling since the purpose for using the model is not related to testing, revising, or evaluating it. This learning approach is the basis of what we characterised as model-based teaching.
- *Learning to revise models* includes devising *new semantic relations between a model and the field of its applicability* and *new model in relation to a new theoretical level*. Here, students modify a given model, but that is not their original mental model. So, although this may be an important stage in learning to model, the extent to which it can be identified as a situation in which students experience modelling as an epistemic scientific practice is quite limited.
- *Learning to reconstruct a model* is *inquiry modelling*. In this case, students really experience all the stages of modelling, but "within the psychological security provided by an awareness that the model is already fully known to and accepted by scientists" (Justi & Gilbert, 2002a, p. 374).
- *Learning to construct a model de novo* and *expressive modelling* are both focused on students constructing their knowledge in the form of a model that is absolutely novel for them.

Although recognising the relevance of studies in which students use or revise models (like Acher et al., 2007; Passmore & Stewart, 2002; etc.), being consistent with the core focus of this book, only studies that involve learning to reconstruct a model or to construct a model *de novo* – that is, those related to what we call modelling-based teaching – are discussed in this chapter.

Modelling-Based Teaching by Reconstructing a Model

As emphasised in the previous section, in many studies students are involved in more than one approach to learning about, or from the use of, models and modelling. In this section, we highlight distinguishing features of studies in which students participate in activities mainly focused on reconstructing a model, by selecting three of them as exemplars.

Assuming that modelling is a non-hierarchical process constituted by the stages of selection, construction, validation, analysis, and deployment (as originally proposed by Hestenes, 1987 and explained in chapter 3), Halloun (1996) involved Lebanese students in the construction of basic models of Newtonian mechanics that would be used to solve problems. In the selection stage, based on the description of the physical situation and guided by the teacher's questions, students were encouraged to select a model from a repertoire of familiar ones. The guided process continued with the teacher asking questions focused on the parameters required for each object, the kinematic concepts required to describe the movement of the object under study, the kinematic laws that describe such movement, etc. The main aim was to help them construct a mathematical model that could be used to solve the problem posed at the beginning of the activity, which is done by constructing (or reproducing) "the composition and the structure of every selected model" (Halloun, 1996, p. 1028). Following (sometimes simultaneously with) the model construction stage, students were guided to check the consistency of their models; to apply them, thus getting answers to the problem being studied (these also have to be interpreted and justified); and to think about and refine their current knowledge from that modelling experience.

In another context – that of an introductory astronomy course – Barab et al. (2000) involved undergraduate students in producing dynamic 3D models of the solar system in order to develop a deep understanding of a series of astronomical phenomena. They used a virtual reality modelling language that simulated a 3D environment on an ordinary computer monitor to answer scientific questions. This was done in the context of an inquiry cycle composed of five steps: analysing and asking questions about a phenomenon highlighted by the instructor (thus generating hypotheses); collecting data related to their previously-encountered hypotheses (from textbooks, the internet, or any other source); incorporating data into a model (running it and, if necessary, by changing some of its parameters); addressing the initial questions (and maybe generating and investigating new ones); and presenting the results. In such a process, students run thought and pseudo-empirical experiments (the latter by using their 3D models), and intensively interact with their peers and the instructor at two distinct moments: during the design and production of the model, and after the presentation of each model. At the end of the process, students wrote individual papers evaluating the models produced and relating them to the characteristics of the real solar system, as well as reflecting on how and in what ways the participation in the process influenced their understanding. According to the authors, this showed their understanding of both the astronomy topics under study and modelling as a way of producing knowledge.

Another interesting study in which students learnt science from reconstructing a model was conducted by Buckley (2000) in a context in which American tenth grade students used an interactive multimedia resource named 'Science for Living: The Circulatory System' (SFL). According to the author,

> SFL was designed to provide access, only on demand, to a wide range of information about the circulatory system and to tolls for manipulating that information to create multimedia presentations. (Buckley, 2000, p. 899).

The main interface screen shows a diagram presenting the key components of the circulatory system and information about the system as a whole. When clicking on parts of the image, each component is highlighted, and additional information (videos, photographs, drawings, as well as lectures and demonstrations presented by experts) about it was provided. Students had to create a multimedia project that explained how digestion and circulation are related and how the food one takes in reach a cell in the finger. After two weeks using STF, students presented their projects, and answered a written test. The learning process of one of the students was analysed in depth, showing that she assembled models of the circulatory system from the many representations of it and from information about internal structures, function, and behaviour of distinct organs that she had accessed from STF. In other words, the elements of STF contributed to the formation and revision of her evolving models of the circulatory system and on how it interacts with organs of the digestive system.

From these studies, we may say that modelling-based teaching by reconstructing a model may support the improvement of students' understanding of both scientific concepts and their epistemological views on models and modelling, as well as the development of their critical reasoning.

Modelling-Based Teaching by Constructing a Model *de novo*

In the literature, there are few MBT proposals that are focused on providing conditions for involving students in constructing a model *de novo*. In this section, we detail and discuss two of them.

The GEM Proposal

Assuming that teaching from a modelling perspective may be difficult for several reasons, Clement (2000) proposed that modelling-based teaching be based on an evolution of students' models. Such a process would start from students' preconceptions, from which they would construct an initial model. Then, the teacher would support students in making small model revisions, that is, in producing a series of intermediate (partial) models until the development of a target model (which, according to him, may not be the model accepted by scientists, but is the one that includes the knowledge that the teacher expects students to acquire at that moment). It seems there was a general agreement around this model-evolution idea among those who have been working on modelling-based teaching, in particular those who approached it from the 'construction of a model *de novo*' perspective (for instance, those from Clement's group, whose main ideas are presented in the book 'Model Based Learning and Instruction in Science' (Clement & Rea-Ramirez, 2008) or in other papers (for example, Khan, 2007; Williams & Clement, 2015), or from other research groups (for instance, Hubber & Tytler, 2013; Justi & Gilbert, 2003; Maia & Justi, 2009b; Mendonça & Justi, 2011; Schwarz & Gwekwerere, 2007; Schwarz et al., 2009)). Some points distinguish Clement's work from these and other research groups, but the main issue is how each group views the process of model evolution.

For Clement's group, the initial student's model is criticised and revised according to the Model Construction Cycle (Fig. 2.1) – also called GEM cycle (for Generation, Evaluation, and Modification of Models). Each of these stages can be fostered by specific kinds of classroom interactions: *the competition mode* (the teacher stimulates students to discuss contradictory models or ideas previously presented); *the disconfirmation mode* (the teacher asks a question that results in the initial model being criticised and abandoned during the discussion); *the modification mode* (the teacher asks a question that results in the initial model being criticised and modified during the discussion); *the accretion mode* (the teacher asks a question and students add a new element of the model at a time) (Clement, 2008b; Rea-Ramirez, Clement, & Núñez-Oviedo, 2008). Independently of the type of classroom interaction, if students have some previous knowledge about the topic under study, they generate many ideas. In other cases, some evaluations and modifications are made by the teacher: when specific content targets are crucial at a given moment, when a modification cannot be obtained from the students, or when the teacher feels that students are ready to receive new information (Clement, 2008a). As a whole, the pattern represents a co-construction process in which both the teacher and the students contribute to ideas and their evaluation.

The empirical studies conducted from this perspective have provided evidence that support such ideas. For instance, Khan (2007) analysed the interactions between a teacher and students in an undergraduate introductory chemistry classroom. In such a context, the instruction was designed to promote students understanding of chemistry at the molecular level through inquiry. The analysis of teacher-students interaction showed successive occurrence of GEM cycles guided by the teacher's actions, and resulting in students' gradual enhancement of their models of molecular structure and understanding of intermolecular forces. This cyclical pattern is also made evident in several chapters of the book (Clement & Rea-Ramirez, 2008), and, in each of them, each one of the specific kinds of classroom interaction is illustrated and discussed. Additionally, according to Núñez-Oviedo, Clement and Rea-Ramirez (2008), the teacher may ask two types of questions: *support questions* and *discrepant questions*. The support questions are generated to activate the student's already existing knowledge in order to relate it to experiences and data provided by modelling activities, thus contributing to the generation and modification of models. On the other hand, the discrepant questions may produce a divergence in relation to the target model and may make students correct their initial model or make additions to it, producing an intermediate model. Such ideas seem interestingly as a way to characterise the co-construction of knowledge in MBT contexts and the influence of specific models on the occurrence of conceptual change (Clement, 2008c). However, in some sense, they may be viewed as a simplification of the process because they imply that, as soon as the teacher asks the 'right' questions and students answer them, the model evolution occurs from one of the situations previously mentioned. This also implies that the teacher's role is the determinant factor in MBT when viewed from this perspective, that is, if the teacher does not have the proper knowledge about the process and skills to conduct it, the desired co-construction of knowledge would not follow. We view this as a serious limitation of this approach, since many studies have shown that teachers content and pedagogical content knowledge on models and modelling are generally far from those necessary to teach from a modelling-based perspective (Crawford & Cullin, 2004; Danusso, Testa, & Vicentini, 2010; Justi & Gilbert, 2002a, 2002b; Justi & van Driel, 2005; Oh & Oh, 2011; Schwarz, 2009).

The Model of Modelling proposal

Basic Principles

By taking into account the limitations of teachers' content knowledge and pedagogical content knowledge on modelling, in the last decade or so, we have been constructing an approach to MBT based on the Model of Modelling diagram (Fig. 2.2). It has been built by focusing on specific issues originated from both empirical data gathered when MBT occurred in classrooms and from a study of the philosophical and cognitive issues on modelling. Several relationships were established and changed in a dynamic ongoing process resulting in the approach that is currently being used. Its main characteristics are described next, and some of them are detailed analysed and exemplified in following chapters.

At the core of our approach is the use of the Model of Modelling diagram in order to basis the design of teaching activities, as well as guidelines regarding the conduct of such activities (that is, suggestions concerning teachers' decision making and actions). This does not mean that students learn about (or even are presented with) the Model of Modelling in order to use it as an algorithm. On the contrary, what is expected is that, as far as students get involved in the modelling activities, they develop a creative way of thinking that include the elements and relationships that compose the Model of Modelling and which could be used in other situations (in scientific contexts or not) (Justi, 2006). In order to support such a development:

- Students must have at least a simple view of models in the scientific context.

  This means that there is no sense in involving them in MBT situations if they view the word 'model' from its ordinary meanings, that is, as copies of objects, beautiful people who pose for photos, types of a given class of objects (cars, for instance), and patterns to be followed. That situation in which students understand models only from their ordinary meanings is very common when students had no previous contact with models in science teaching. Although this seems a fictional situation for those from countries where the epistemic practices of science are introduced to children in primary schools, it is very common in many countries where science is mainly taught from a traditional approach, that is, by focusing on the transmission of non-contextualised content. For instance, in all studies we conducted in Brazil, the initial activity aimed at helping students understand that, in the scientific context:

- models are not copies of the entity being modelled, patterns to be followed, or distinct types of a given class of objects;
- models can be used for many functions, mainly to simplify complex entities, to represent such an entity, to support explanations and predictions, etc.;
- models can be changed;
- it is possible to have more than one model for a given entity.

We recognise that even this simple view on models may be difficult to be fully grasped by some students. Our empirical data (for instance, those from the studies reported in Maia & Justi, 2009b; Mendonça & Justi, 2011; and many others unpublished works) shows that students' participation in such an initial activity sows the seeds of a comprehensive understanding of models, which continued growing as far as students are involved in the modelling activities.

- The teacher has to motivate students to participate in the activities.

  In order to do so, two major conditions must be satisfied. First, in a regular teaching situation in which students are going to learn something from MBT, it is the teacher's responsibility to initiate the process, that is, to propose the aims for producing a model. However, it is essential that this were done in a context that makes sense to students. In other words, if students neither attribute meaning to the teacher's request nor understand the advantages of producing a model in a given context, it is unlikely that they will make any effort to engage with the activity. Second, students' ideas and doubts expressed during the process have to be valued and respectfully treated. We view this as essential if we wish to involve students cognitively and emotionally in the modelling activities. The shaping of such a respectful environment may contribute to the development of students' skills of: accepting the existence of contradictions, inconsistencies and similarities between distinct ideas; analysing someone's else ideas in a supportively critical way; accepting and valuing criticisms – all of which are essential not only to perform scientific practices like modelling, but also, and mainly, to act as a citizen in present-day societies.
- Students should work in small groups in order to help each other, to discuss doubts, to try to produce a group consensus model to be communicated to, and discussed with, the whole class.

  As students have idiosyncratic skills, previous knowledge and experiences, their involvement in a challenged activity may support real collaborative work and approximate it to authentic scientific practice.

In order to support science teaching from our MBT approach, for each given curricular topic to be taught, we produced a sequence of activities that support students' engagement in modelling by making them experience all the stages of the process.[2] This engagement should be carefully planned and scaffolded if it is to help students construct knowledge during the process. This means that the activities have to guide students to think about specific issues or to perform certain experiments and, simultaneously, be as 'open' as possible, that is, they should not ask for a uniquely right answer. This will allow students to be creative so that they feel that they could ask questions, propose tests, make and evaluate explanations and predictions on their own. This balance may be difficult to be achieved, but its existence in the activities is essential to their success, particularly in cases where teachers are also learning about modelling and how to teach from a MBT approach.

In the previous chapter, we characterise each of the stages of modelling according to the Model of Modelling framework. Such a characterisation builds on both some of our initial theoretical discussions on the framework (Justi, 2006; Justi & Gilbert, 2002a) and evidence gathered in many empirical studies conducted in the last decade. From them, we identified the skills and abilities[3] that one needs to perform each of the stages of modelling (Table 4.1).

Taking into account the basic principles discussed in this sub-section, the characterization of each of the stages of modelling according to the Model of Modelling framework (discussed in Chap. 3), and the skills and abilities needed to perform each of the stages (Table 4.1), we have developed some guidelines for designing and conducting the modelling activities. They are described next.

The Design and Conduct of the Teaching Sequences

The starting point for students' engagement in modelling activities is the understanding of the aims of the model to be produced (or, at least, an initial partial understanding of them). This means that students ought to understand the question(s) that they will seek to answer. Following this, the first activities ought to involve students in having experiences with the entity to be modelled. This may mean that they have:

- to remember previous ideas (either acquired in school situations or in ordinary, everyday, ones);
- to acquire information from external sources (textbooks, the internet, etc.); and
- to gather information by performing practical work and/or simulations.

---

[2] Although the model applies to the whole of sciences, so far we have been able to use it in chemistry classes.

[3] The distinction between ability and skill is subtle and, in some sense, controversial in the education literature. In this book, we assume abilities to be the generic, non-quantifiable quality of being able to do something, and skills as a composite of abilities, techniques and knowledge (proficiencies) developed through training or experience and that makes one does tasks at a higher degree or standard. So, abilities help people to develop their skills.

Table 4.1 Skills and abilities related to each of the stages of modelling according to the Model of Modelling framework

Stages of modelling	Related skills and abilities
Creation of a proto-model	A. To define the aims of the model or to understand the aims proposed by someone else
	B. To search for information about the entity to be modelled (in the modeller's cognitive structure and/or from external sources: bibliography, empirical activities, etc.)
	C. To select and organise the available information by having the aims of the model in mind
	D. To use analogical reasoning or a mathematical tool on which to base the proto-model
	E. To integrate all the previous information and experiences into the creation of a proto-model
Expression of the proto-model ⇨ production of the artefact (model)	F. To visualise the proto-model
	G. To use any of the models of representation (or a combination of them) to express the proto-model
	H. To adapt the expressed to the mental proto-model (or vice-versa) by making comparisons between each other or between them and the aims of the model
Test of the model	I. To plan and conduct thought experiments
	J. To plan and conduct empirical experiments
	K. To critically analyse the results of the tests
	L. To modify or to reject the model
Evaluation of the model	M. To identify the limitations of the model by trying to use it in different contexts
	N. To identify the scope of the model by trying to use it in different contexts
	O. To convince others of the validity of the model
	P. To persuade others to use a given model

The use of practical work as a source of information for students is particularly relevant when they are modelling abstract entities that are not evident in ordinary situations – as always occur in chemistry – and in which they have to count on interpretations of the behaviour of such entities to produce their models. But, independent of the type of entity being modelled, the most important characteristic of the practical activities involved in MBT is that they have to be investigative, that is, they have to provide students with opportunities to think, to use their previous knowledge and experiences to solve a problem whose answer is not obvious to them (Duggan & Gott, 1995). Thus, in the context of our MBT approach, experiments cannot be conducted only to collect data that might be obtained from other sources. Rather than this, they have to challenge students to think about the data, the context in which they are gathered, their previous ideas that may explain the observations, and the meaning they attribute to the data in that and other related contexts. In this way, the involvement in such activities may support the establishment of the necessary relationships between the data and the entity being modelled. In this sense, the questions included in the experimental activity or additional ones asked by the teacher are of pivotal importance.

The initial activities (that will culminate in the production of the students' initial proto-model) should also guide the selection of an origin for the model. This is one of the modelling steps in which analogical reasoning assumes an important role. Here, analogies may be drawn either spontaneously, that is, as part of the students' reasoning process, or from teacher's questions that make students produce relational comparisons between two domains. However, such teachers' questions have to be originated from ideas expressed by the students at the time. Otherwise, the resulted analogy would be imposed by the teacher and wouldn't likely contribute to the production of students' proto-models. Due to both the fact that the selection of the origin for the model is basically a mental process, and the fact that it is difficult to predict students' ideas and ways of reasoning, it is almost impossible to include questions concerning the selection of the origin to the model in the modelling activities.

Having producing an initial proto-model, students have to express it in order to communicate and/or discuss their ideas. This may be done by using one or more of the most common modes of representation: verbal (oral or written), 2D (drawings, schemes, graphics, etc.), 3D (from the use of any available concrete material), gestural, virtual, and mathematical. This stage is highlighted in the Model of Modelling framework as a distinct one because it involves cognitive processes distinct from those used when the initial proto-model is produced, mainly those concerning the transposition from a mental visualisation of the model to its external representation. As discussed in Chaps. 3 and 7, we view this transposition as essential because it is from one's interaction with the expressed models that they can perform the epistemic practices through which they are produced.

Modelling activities can support students' experience of this stage by requesting them not only to express their models but also to explain why a given mode of representation is being used, and the meaning of the codes of representation used in each of them. This kind of explanation is important because it may support students' understanding that each mode enables the representation of specific elements, as well as that the use of each of them requires distinct cognitive operations.

When requesting students to express their models, the activities can either identify a specific mode of representation to be used or not, that is, students can just be asked to express their models. The decision about specifying the mode of representation would be mainly guided by the levels of students' expected learning. However, both options have advantages and disadvantages. When the mode of representation is identified, the activity supports the subsequent comparative analysis of both distinct models and specific codes of representation used (maybe in different ways) by each group of students. This may be very useful when students are novices at modelling. On the other hand, the teacher would need to explain that:

- that given mode of representation was chosen due to some facility which it has in relation to the aim of the model, not because it is the "right" way to represent it; and
- other modes could be used to provide distinct representations for the same model, which would be more, or less, useful in diverse contexts.

When the activity does not request the use of a specific mode of representation, students have the opportunity to think about and to make decisions from their previous modelling experiences and in respect of the aims of the particular model being produced. This would require that students have familiarity with some modes of representation, and be creative in using them – which may be difficult for novices. For instance, once one of us faced a group of secondary students who had no idea about how to represent a chemical structure. As some of their colleagues were doing this by using polystyrene balls, the teacher suggested that they should do the same. They seemed surprised by the teacher's suggestion because they have never thought about representing an abstract entity (whose meaning they have just started to understand) by using a material object. On the other hand, the freedom to select the mode of representation to be used may also contribute to the simultaneous occurrence of the stages of creation and expression of the model, since specific outcomes in the expression of the model could be the base for the emergence of ideas that, in turn, could enhance or completely change the initial proto-model(s) (Justi, 2006).

In order to be consistent with such suggestions, teachers' actions concerning the expression of the students' models ought to be focused on:

- explaining about the modes of representations, when students do not have familiarity with those that might be used by them; and
- supporting students' clear communication of their models.

Teachers should thus lay emphasis on the questions asked in the written activities or ask other questions about (i) the modes of representations used by students, (ii) the ways students used them, or (iii) the ideas embedded in their models. Additionally, students should be given opportunities to communicate their expressed model to the whole class. By mediating this moment, the teacher may encourage initial discussions about distinct models. This is the first time that students try to defend their ideas. Our personal experience of analysing data from many classes shows that this is always a valuable moment for students because, whilst defending their ideas, students increase their awareness and understanding of some of the details involved in such ideas. Additionally, they start to think about new ideas from the analysis of their colleagues' models.

In respect of the testing stage, students may design and conduct tests for their models. However, at least initially, this may be difficult for students who are constructing a model *de novo* and, at the same time, learning the major concepts involved in that model. As all our sequences of activities are produced for teaching specific topics for students who have never studied them before, we always plan some activities that suggest tests for students' models (to which their own suggested tests can be added). As this stage aims at checking whether the current model fulfils its purposes and, if necessary, modifying the model or rejecting it and restarting the process, it is very useful to have multiple tests available. The existence of multiple tests is also important because, as the groups of students produce models independently of each other, their models may be entirely distinct. Therefore, the interpretation of the result(s) of a given test in light of each model may imply in some models be modified more intensively than others before the students concluded that their model fulfil its purposes. When multiple tests are provided, this process is facilitated because each of them may either address a specific flaw in students' models or support the reinterpretation of the available information in a distinct way. For instance, in the MBT sequence for ionic bonding, students are asked to produce a model for 'cooking salt dissolved in water' and 'the salt plus water system after the evaporation of water'. In all classes, this request results in the production of a 'NaCl molecule' model, that is, a model comprise by two bound balls (each one representing one of the ions). In the following activities, students have opportunities to test their models twice. They are asked to use their model to explain: the high melting temperature of sodium chloride, and why the empirical value of energy liberated when the actual sodium chloride is formed is higher than the value liberated when the pair $Na^{+}Cl^{-}$ is formed; and why minerals that are constituted in exactly the same way as the sodium chloride are friable and can be cleave along specific planes. Some students interpret the sodium chloride's high melting point as a piece of evidence of the existence of several bonds and propose a new model comprised of a lattice of ions. With this new model, they are able to produce a reasonable explanation for the energy values in the following test. But most of the students interpret the sodium chloride's high melting point as a piece of evidence that 'the' bond that binds the ions of the 'NaCl molecule' is strong. Those students only start to realise that their model needs to be modified when they perform the following activity (the second test) (Mendonça & Justi, 2011).

Additionally, the evidence provided by the tests should not only show if the model is flawed. They can also provide additional information to support the modification of that model or the production of a completely new one. Thus, in order to plan such testing activities, it is necessary to take into account:

- the literature on students' ideas on that given topic, from which we can learn about students' main difficulties;
- previous teaching experiences in which the topic have been discussed, in order to identify questions that may support students discussion of their previous ideas.

Both approaches would allow the identification of those ideas that most often should be changed in students' models (or added to them) so that they learn the curricular model. The identification of such ideas is also important in order to define whether the test is going to be designed involving an empirical and/or a thought experiment. In the teaching sequence mentioned in the previous paragraph, those specific tests were designed because the authors were sure (from the literature on students' misconceptions and their previous teaching experiences) that students always initially think that the structure of sodium chloride is that of a 'molecule'. In that case, their previous teaching experiences also supported the identification of questions to be asked to students during the activities (for instance, 'what is a molecule?'; 'when a positive ion attracts a negative one, do they neutralise each other?'; 'how many positive ions can a negative one attract, and vice versa?').

One significant feature of the activities that involve tests for the models is that the questions that are posed to students have to support the students' own interpretation of the results obtained from those tests. This is regarded as crucial in the enhancement of students' skills in creativity and in their critical analysis of their models, in the design of other tests, and in formulating their own questions. Additionally, in the case of empirical tests, the degree to which they (and the included questions) are 'open' (that is, do not require a 'right' or 'unique' answer) may contribute to students' seeing that conducting a practical work in modelling activities is not 'following a recipe'.

Data from the MBT studies based on the Model of Modelling also shows that, when students are testing their models, teachers' actions are focused on:

- supporting the development of students' skills related to the planning and conducting of thought experiments, especially in encouraging them to be creative when doing so;
- helping students to analyse the results of the tests by evaluating them in respect of the meanings they attribute to the work so far, and the relationships that they are establishing between the results and their previous models.

For instance, teachers conducting the modelling-based activities on ionic bonding are advised to continuously ask students (i) about what they are thinking when performing the test activities, (ii) to justify their ideas, (iii) to think in other properties of sodium chloride and to try to explain them using their models. When students have already changed their models, teachers are also advised to ask them to use both models (the previous and the current one) in each situation in order to compare their explanatory power in all situations.

The teacher's actions may also provide counter-examples that support students' understanding of a given result or relationship. For instance, when students are in doubt about the meaning of 'a high melting point', the teacher may provide them with melting points of other ordinary substances that are liquid at the room temperature and ask them to explain why they exist in that physical state whilst other compounds (like sodium chloride) are solid.

As in other moments in the work, the level of involvement of the teacher in students' reasoning during the testing of the models will vary depending on the students' previous experiences with modelling activities, their previous knowledge about the entity that is being modelled, and the teacher's aims in terms of the development of students' knowledge and skills. For instance, in one study conducted from the Model of Modelling framework, all lessons in a class were observed and registered in video for several months. As those students have never had any experience with modelling activities before, it became clear that, as they experienced sequential modelling activities, they became more autonomous, and the teacher had to help them less often (Mendonça & Justi, 2013).

When students are testing their models, the teacher must choose an appropriate moment to organise a whole class discussion. At such times, each group may have better (more complex) arguments to both clearly explain their models and try to convince their colleagues of the validity and explanatory power of their models. From discussions and recognition of the relative validity of individual models, one or more class-consensus model(s) may be produced. In this process, it is essential that the teacher does not evaluate each group model as correct or wrong, but that s/he emphases, or makes students realise, how elements from individual models contribute to fulfil a given purpose for the required model. However, during such discussions, the teacher may realise that essential elements of the curricular model are not part of any of the students' models. This is an excellent opportunity for the teacher to show that MBT is not consistent with an authoritarian imposition of scientific knowledge. On the contrary, teachers are expected to present such elements as other alternatives to be analysed by students – which may contribute to integrate at least part of such elements into their models (Justi, 2006, 2009).

After reaching an agreement around one or more class-consensus model(s), the final activities of each sequence is to provide opportunities for using it (them) in different and distinct contexts.[4] Consequently, the scope and limitations of the model(s) have to be identified and discussed. In the teaching sequence for ionic bonding, for instance, students are asked to use their models to explain the behaviour and properties of other compounds that are constituted of ionic lattices.

The evaluation of the 'final' models is another opportunity for students to try to persuade their peers of the comprehensiveness of a given model. Students tend to produce complex and strong argument in favour of their models when they conclude that they are able to explain situations in distinct contexts and/or which have few single limitations.

As a result of such discussions, a class-consensus model is produced. But, as it does not necessarily correspond to the curricular model, the teacher may help students to narrow the distance between such models. This may be done by questioning students, supporting the establishment of additional relationships, or providing additional information. For instance, in all cases when we used modelling activities to teach ionic bonding, the class consensus model produced after the last activity included the structure of ionic compounds as a lattice. However, such lattices had no specific shapes because those students (who were learning ionic bonding for the first time) could not interpret X-Ray spectra in order to decide about the accepted arrange of the ions. So, the teacher emphasised the importance of representing ions in a lattice (as they did in their models), but informed them about the existence of technologies that allow scientists to decide about the shape of the lattice. Students viewed this information just as a contribution to the model that they have created and had no problem in accepting and incorporating it into their models (Mendonça & Justi, 2011).

Outcomes from MBT According to the Model of Modelling Framework

From a series of investigations on the use of modelling activities based on the Model of Modelling framework conducted in distinct classes, we realised that most of the students:

[4] Although we recognise that modelling is a continuous process, in the educational context the curriculum and other school constraints restrict the initial and explicit teaching of each topic to a specific period of time. Thus, the design of the activities have to aim at contributing to the building of students' models related to a given curricular model.

- more effectively learnt the chemical topics that were related to the models that they had constructed. For instance, the study in which the MBT for chemical equilibrium was analysed (Maia & Justi, 2009b) was conducted with 14–15 year-old students when they were in the first grade of the medium level school (a school level in Brazil that is constituted by three academic years, and is attended by students from 14 to 15 year-old). Two years later, that is, when those students were in the third grade, their chemistry teacher was one of the members of the same research group. When revising all the chemistry topics taught during the first and second grades before their university entrance exams, the teacher realised that chemical equilibrium was the one with which they had less difficulties. This was so because most of them used the ideas discussed in the MBT activities and that constituted their models of chemical equilibrium to appropriately solved all problem situations presented by the teacher, that is, without expressing any of the common alternative conceptions related to the concept;
- developed a broad understanding about the nature and uses of models, and about the importance of modelling in the construction of scientific knowledge. For instance, in one of the studies in which data was collected for several months (during which three teaching sequences based on modelling were used), both the development of students' investigative skills during the process and their views on models and modelling were investigated (Maia, 2009; Maia & Justi, 2009a). In that study, all the lessons were video recorded, and students were interviewed three times. From the analysis of the videos, the authors identified how students performed each of the stages of modelling in sequential activities. Then, in the interviews they were asked to justify their actions and, from their answers, they were questioned about their views on models and modelling (including the role of modelling in the development of scientific knowledge). Such a set of data supported the authors' conclusions that, as students participated in more modelling activities, they broadened their views on:
  - models, from something that only represents an entity to something that (i) has limitations, (ii) can be changed, and (iii) can be used to explain specific aspects (that is, assuming the existence of multiple models for a given entity);
  - modelling, by (i) clearly understanding the importance and role of having experiences with the modelled entity; (ii) clearly expressing, discussing, and modifying their models; and (iii) using models in different contexts.

We also realised that, when the teacher had a deeper understanding about scientific epistemic practices (like modelling, argumentation, visualization, investigation) as well as about science (its nature, mainly in terms of cognitive, philosophical and historical views), students also developed skills related to such practices. Evidence that supports this affirmative is provided and discussed in following chapters.

The premise that participating in the practice of modelling may result in students' acquisition of knowledge about, and skills related to, modelling is also present in the work of other researchers. For instance, according to Lehrer and Schauble (2012), due to the complexity of modelling, it develops "only over the long term, both historically within the sciences and ontogenically within the lifetime of individuals" (Lehrer & Schauble, 2012, p. 702). From the recognition that educational materials and curricula do not generally address the complexity of modelling, they acknowledge that

> modelling practices need to be systematically and cumulatively built and fostered, rather than simply being injected at high school or university or, even worse, taken for granted as self-evident (Lehrer & Schauble, 2012, p. 702).

Therefore they propose the structure of a learning progression to understand the development of modelling to be used since kindergarten to support students' participation in modelling practices and construction of knowledge. More details about their proposal of a learning progression are discussed in Chap. 10, where we also propose a learning progression based on theoretical and empirical data discussed in the next chapters.

Concluding Remarks

MBT based on the Model of Modelling framework is characterised here as an approach for teaching basic curricular concepts when students are learning them for the first time – and, on many occasions, without any previous modelling experience. The results of the studies conducted in classes where some concepts were taught from this MBT approach in the same academic year can be used to project that, when students had built up their knowledge and skills on modelling and on some topics, they will be able to participate in what have been named 'authentic modelling practices' (Prins et al., 2008). Those practices would occur in selected contexts in which students could use modelling to analyse or to solve real-world or socio-scientific problems. The complexity of the challenge justifies its non-inclusion in science education for novices. In the Netherlands, for instance, Prins, Bulte, and Pilot (2011) report good results when 16–17 years-old upper secondary education students participated in an authentic chemical practice on 'modelling drinking water treatment'. On the other hand, depending on the selected context, the problem that is posed to students, and their prior related models, it is likely that the modelling practices performed by students in these contexts would be characterised as 'revising models' or 'reconstructing a model'.

In sum, notwithstanding the nature of the topic and the entity to be modelled, the studies based on teaching approaches that involve students in *constructing a model de novo* have shown that, when adequately conducted, they usually contribute to students' learning. Additionally, the MBT approach based on the Model of Modelling framework addresses the complexity of modelling by making students experience and reflect on each of its stages. At the same time, our proposal is based on the premise that modelling practices have to be systematically and cumulatively built. As emphasised by Lehrer and Schauble (2010), the building of modelling practices in this way is essential rather than by just including a sporadic modelling activity "or, even worse, taken for granted as self-evident" (Lehrer & Schauble, 2010; 2012, p. 702). Therefore, the use of this approach can characterise an educational reality completely different from those commonly practiced in many science education classrooms.

In the introductory section of this chapter, we claim that MBT can support authentic science education contexts because, among other reasons, it can provide conditions for students' engagement in scientific epistemic practices, that is, in specific ways scientists "propose, justify, evaluate, and legitimize" scientific knowledge claims (Kelly, 2008, p. 99). The following discussion on MBT by construction a model *de novo* shows that 'developing and using models' is not a single scientific practice, as it may be understood from the list of practices for K-12 science classrooms proposed in the most recent American science education policy document (National Research National Research Council, 2012).[5] When students participate in a teaching sequence whose activities are based on the Model of Modelling framework, for instance, they may have opportunity to perform not only those eight practices, but also other relevant scientific practices (like visualisation) and types of reasoning (like the analogical one). In fact, from both the characterisation of the modelling activities from this perspective (discussed in a previous section of this chapter) and the identification of the skills and abilities related to the performance of each of the stages of modelling (Table 4.1), it becomes clear that the unique practice that is not intrinsically involved in any of the modelling stages is 'using mathematical and computational thinking'. Its involvement in all the stages depends on both the topic being modelled (for instance, it may suit some physics topics) and the students' intellectual level (since younger ones tend to study science only from a qualitative perspective). Therefore, MBT by constructing a model *de novo* implies in providing conditions for students: (i) become members of a social community of knowledge building; (ii) learn broader and specific scientific ideas; (iii) develop their understanding of the epistemic foundation of science (how we know what we know); (iv) develop their meta-knowledge of science as practices and products integrated by human beings' creative reasoning in dynamic and complex ways. These outcomes of science education have been emphasised as essential to improve scientific literacy of twenty-first century citizens (for instance, in Allchin, 2014; Duschl & Jiménez-Aleixandre, 2012; Hodson, 2014b; Millar, 2006; Osborne, 2014; Windschitl et al., 2008).

Finally, students involved in constructing a model *de novo* can also understand that, besides representing, models have many other functions, including a generative constructive one, that is, they can be used to support the construction of other models and the generation of new systems to be studied (Peschard, 2011). In other words, students can then view models as epistemic artefacts. This may launch new

[5] The list comprises eight practices, namely: "asking questions; developing and using models; planning and carrying out investigations; analyzing and interpreting data; using mathematics and computational thinking; constructing explanations; engaging in argument from evidence; obtaining, evaluating, and communicating information" (National Research National Research Council, 2012, p. 42).

challenges to students; and create new reasons to learn science, and new ways to think in school or ordinary context involving science.

References

Acher, A., Arcà, M., & Sanmartí, N. (2007). Modeling as a teaching learning process for understanding materials: A case study in primary education. *Science Education, 91*(3), 398–418.

Allchin, D. (2014). From science studies to scientific literacy: A view from the classroom. *Science & Education, 23*(9), 1911–1932.

Barab, S. A., Hay, K. E., Barnett, M., & Keating, T. (2000). Virtual solar system project: Building understanding through model building. *Journal of Research in Science Teaching, 37*(7), 719–756.

Buckley, B. C. (2000). Interactive multimedia and model-based learning in biology. *International Journal of Science Education, 22*(9), 895–935.

Campbell, T., Oh, P. S., & Neilson, D. (2012). Discursive modes and their pedagogical functions in model-based inquiry (MBI) classrooms. *International Journal of Science Education, 34*(15), 2393–2419.

Clement, J. J. (1989). Learning via model construction and criticism: Protocol evidence on sources of creativity in science. In J. A. Glover, R. R. Ronning, & C. R. Reynolds (Eds.), *Handbook of creativity* (pp. 341–381). New York, NY: Plenum.

Clement, J. J. (2000). Model based learning as a key research area for science education. *International Journal of Science Education, 22*(9), 1041–1053.

Clement, J. J. (2008a). *Creative model construction in scientists and students – The role of imagery, analogy, and mental simulation*. Dordrecht, The Netherlands: Springer.

Clement, J. J. (2008b). Model based learning and instruction in science. In J. J. Clement & M. A. Rea-Ramirez (Eds.), *Model based learning and instruction in science* (pp. 1–9). Dordrecht, The Netherlands: Springer.

Clement, J. J. (2008c). The role of explanatory models in teaching for conceptual change. In S. Vosniadou (Ed.), *International handbook of research on conceptual change* (pp. 417–452). New York, NY/London, UK: Routledge.

Clement, J. J., & Rea-Ramirez, M. A. (2008). *Model based learning and instruction in science*. Dordrecht, The Netherlands: Springer.

Crawford, B. A., & Cullin, M. J. (2004). Supporting prospective teachers' conceptions of modeling in science. *International Journal of Science Education, 26*(11), 1379–1401.

Danusso, L., Testa, I., & Vicentini, M. (2010). Improving prospective teachers' knowledge about scientific models and modelling: Design and evaluation of a teacher educational intervention. *International Journal of Science Education, 32*(7), 871–905.

Duggan, S., & Gott, R. (1995). The place of investigations in practical work in the UK National Curriculum for Science. *International Journal of Science Education, 17*(2), 137–147.

Duschl, R., & Jiménez-Aleixandre, M. P. (2012). Epistemic foundations for conceptual change. In S. M. Carver & J. Shrager (Eds.), *The journey from child to scientist: Integrating cognitive development and the education sciences* (pp. 245–262). Washington, DC: American Psychological Association.

Gilbert, J. K. (2004). Models and modelling: Routes to a more authentic science education. *International Journal of Science and Mathematics Education, 2*, 115–130.

Gobert, J. D., & Buckley, B. C. (2000). Introduction to model-based teaching and learning in science education. *International Journal of Science Education, 22*(9), 891–894.

Halloun, I. A. (1996). Schematic modeling for meaningful learning of physics. *Journal of Research in Science Teaching, 33*(9), 1019–1041.

Halloun, I. A. (2007). Mediated modeling in science education. *Science & Education, 16*(7 & 8), 653–697.

Hestenes, D. (1987). Toward a modeling theory of physics instruction. *American Journal of Physics, 55*(5), 440–454.
Hodson, D. (1992). In search of a meaningful relationship: An exploration of some issues relating to integration in science and science education. *International Journal of Science Education, 14*(5), 541–562.
Hodson, D. (2014a). Learning science, learning about science, doing science: Different goals demand different learning methods. *International Journal of Science Education, 36*(15), 2534–2553.
Hodson, D. (2014b). Nature of science in the science curriculum: Origin, development, implications and shifting emphases. In M. Matthews (Ed.), *International handbook of research in history, philosophy and science teaching* (pp. 911–970). Dordrecht, The Netherlands: Springer.
Hubber, P., & Tytler, R. (2013). Models and learning science. In R. Tytler, V. Prain, P. Hubber, & B. Waldrip (Eds.), *Constructing representations to learn in science* (pp. 109–134). Rotterdam, The Netherlands: Sense.
Justi, R. (2006). La Enseñanza de Ciencias Basada en la Elaboraci6n de Modelos [Modelling-based Science Teaching]. *Enseñanza de las Ciencias, 24*(2), 173–184.
Justi, R. (2009). Learning how to model in science classroom: Key teacher's role in supporting the development of students' modelling skills. *Educacion Quimica, 20*(1), 32–40.
Justi, R., & Gilbert, J. K. (2002a). Modelling, teachers' views on the nature of modelling, implications for the education of modellers. *International Journal of Science Education, 24*(4), 369–387.
Justi, R., & Gilbert, J. K. (2002b). Science teachers' knowledge about and attitudes towards the use of models and modelling in learning science. *International Journal of Science Education, 24*(12), 1273–1292.
Justi, R., & Gilbert, J. K. (2003). Models and modelling in chemical education. In J. K. Gilbert, O. d. Jong, R. Justi, D. F. Treagust, & J. H. V. Driel (Eds.), *Chemical education: Towards research-based practice* (pp. 47–68). Dordrecht, The Netherlands: Kluwer.
Justi, R., & van Driel, J. (2005). The development of science teachers' knowledge on models and modelling: Promoting, characterizing and understanding the process. *International Journal of Science Education, 27*(5), 549–573.
Kelly, G. (2008). Inquiry, activity, and epistemic practice. In R. Duschl & R. E. Grandy (Eds.), *Teaching scientific inquiry: Recommendations for research and implementation* (pp. 99–117). Rotterdam, The Netherlands: Sense.
Khan, S. (2007). Model-based inquiries in chemistry. *Science Education, 91*(6), 877–905.
Lehrer, R., & Schauble, L. (2010). What kind of explanation is a model? In M. K. Stein (Ed.), *Instructional explanations in the disciplines* (pp. 9–22). New York, NY: Springer.
Lehrer, R., & Schauble, L. (2012). Seeding evolutionary thinking by engaging children in modeling its foundations. *Science Education, 96*(4), 701–724.
Maia, P. F. (2009). *Habilidades Investigativas no Ensino Fundamentado em Modelagem [Investigative Skills in Modelling-based Teaching].* PhD thesis, Universidade Federal de Minas Gerais, Belo Horizonte, Brazil.
Maia, P. F., & Justi, R. (2009a). Desenvolvimento de habilidades em atividades de modelagem (The development of investigative skills in modelling-based teaching contexts). *Enseñanza de las Ciencias, 27*(extra), 776–779.
Maia, P. F., & Justi, R. (2009b). Learning of chemical equilibrium through modelling-based teaching. *International Journal of Science Education, 31*(5), 603–630.
Mellar, H., Bliss, J., Bliss, J., & Boohan, R. (1994). Introduction: Modelling and education. In H. Mellar, J. Ogborn, & C. Tompsett (Eds.), *Learning with artificial worlds: Computer based modelling in the curriculum* (pp. 1–7). London, UK/Washington, DC: The Falmer Press.
Mendonça, P. C. C., & Justi, R. (2011). Contributions of the Model of Modelling diagram to the learning of ionic bonding: Analysis of a case study. *Research in Science Education, 41*(4), 479–503.
Mendonça, P. C. C., & Justi, R. (2013). The relationships between modelling and argumentation from the perspective of the Model of Modelling diagram. *International Journal of Science Education, 35*(14), 2007–2034.

Millar, R. (2006). Twenty first century science: Insights from the design and implementation of a scientific literacy approach in school science. *International Journal of Science Education, 28*(13), 1499–1521.

Morgan, M. S. (1999). Learning from models. In M. S. Morgan & M. Morrison (Eds.), *Models as mediators – Perspectives on natural and social science* (pp. 347–388). Cambridge, UK: Cambridge University Press.

National Research Council. (2012). *A framework for K-12 science education: Practices, crosscutting concepts, and core ideas*. Washington, DC: The National Academies Press.

Núñez-Oviedo, M. C., Clement, J. J., & Rea-Ramirez, M. A. (2008). Developing complex mental models in biology through model evolution. In J. J. Clement & M. A. Rea-Ramirez (Eds.), *Model based learning and instruction in science* (pp. 173–193). Dordrecht, The Netherlands: Springer.

Oh, P. S., & Oh, S. J. (2011). What teachers of science need to know about models: An overview. *International Journal of Science Education, 33*(8), 1109–1130.

Osborne, J. (2014). Teaching scientific practices: Meeting the challenge of change. *Journal of Science Teacher Education, 25*(2), 177–196.

Passmore, C. M., & Stewart, J. (2002). A modeling approach to teaching evolutionary biology in high schools. *Journal of Research in Science Teaching, 39*(3), 185–204.

Peschard, I. (2011). Making sense of modelling: Beyond representation. *European Journal for the Philosophy of Science, 1*(3), 335–352.

Prins, G. T., Bulte, A. M. W., van Driel, J., & Pilot, A. (2008). Selection of authentic modelling practices as contexts for chemistry education. *International Journal of Science Education, 30*(14), 1867–1890.

Prins, G. T., Bulte, A. M. W., & Pilot, A. (2011). Evaluation of a design principle for fostering students' epistemological views on models and modelling using authentic practices as contexts for learning in chemistry education. *International Journal of Science Education, 33*(11), 1539–1569.

Rea-Ramirez, M. A., Clement, J. J., & Núñez-Oviedo, M. C. (2008). An instructional model derived from model construction and criticism theory. In J. J. Clement & M. A. Rea-Ramirez (Eds.), *Model based learning and instruction in science* (pp. 23–43). Dordrecht, The Netherlands: Springer.

Schwarz, C. V. (2009). Developing preservice elementary teachers' knowledge and practices through modeling-centered scientific inquiry. *Science Education, 93*(4), 720–744.

Schwarz, C. V., & Gwekwerere, Y. N. (2007). Using a guided inquiry and modeling instructional framework (EIMA) to support preservice K-8 science teaching. *Science Education, 91*(1), 158–186.

Schwarz, C. V., Reiser, B. J., Davis, E. A., Kenyon, L., Achér, A., Fortus, D., … Krajcik, J. (2009). Developing a learning progression for scientific modeling: Making scientific modeling accessible and meaningful for learners. *Journal of Research in Science Teaching, 46*(6), 632–654.

Svoboda, J., & Passmore, C. M. (2013). The strategies of modeling in biology education. *Science & Education, 22*(1), 119–142.

Tiberghien, A. (1994). Modeling as a basis for analyzing teaching-learning situations. *Learning and Instruction, 4*(1), 71–87.

van Joolingen, W. (2004). *Roles of modeling in inquiry learning*. Paper presented at the IEEE International Conference on Advanced Learning Technologies, Joensuu, Finland.

Williams, E. G., & Clement, J. J. (2015). Identifying multiple levels of discussion-based teaching strategies for constructing scientific models. *International Journal of Science Education, 37*(1), 82–107.

Windschitl, M., Thompson, J., & Braaten, M. (2008). Beyond the scientific method: Model-based inquiry as a new paradigm of preference for school science investigations. *Science Education, 92*(5), 941–967.

Chapter 5
Learning Scientific Concepts from Modelling-Based Teaching

Abstract Whilst teaching for modelling competence is only gradually being established in schools, teaching for concept development is firmly established. If the best use is to be made of curriculum time, the relationship between the meanings of the two activities must be examined and cognitive efficiencies sought. The conventional meaning of 'concept' enables the notions of 'conceptual formation', 'conceptual evolution' and 'conceptual change' to be discussed in respect of single concepts. Science education in respect of concepts involves appropriate changes in a person's ontology, epistemology, and meta-representational competence. However, the established model of conceptual change meets a series of problems arising because these three conditions are not all met. If an artefactual view of 'concept' is adopted, the learning approaches embedded in MBT can be adopted in single-concept work and these problems overcome. Thus the meanings of concept and model coalesce.

The words 'concept' and 'model' are often used somewhat interchangeably both in general conversation and in science and science education. Exploring the relationship between the two is necessary if the relationship between 'modelling-based teaching' and the long-established field of 'teaching for concept development' is to be clarified.

The Relationship Between Concepts and Models

The meanings of 'concept' and 'model' are obscured by the often imprecise use of words in connection with them in the literature. We will try to avoid that pitfall by focusing on definitions and by analysing the differences between the two words. A good place to start is with a definition of 'definition' itself: a definition is the exact form of socially accepted words in which the precise meaning of a particular word is given. The use of 'socially-acceptance' is very important here, for meanings, and hence definitions, alter as word-use changes over a period of time, even within the community of scientists, as is reflected in the notion of 'paradigm shift' (Kuhn, 1996).

J.K. Gilbert, R. Justi, *Modelling-based Teaching in Science Education*, Models and Modeling in Science Education 9, DOI 10.1007/978-3-319-29039-3_5

The Definition of 'Concept' in Science and Science Education

As Medin and Rips (2005) point out in their comprehensive review of the complex literature on the meaning of 'concepts':

> The concept of concepts is difficult to define, but nobody doubts that concepts are fundamental to mental life and human communication. (Medin & Rips, 2005, p. 37)

That is to say: 'the idea is important, whatever it is'!

Their view, in 2005, was that there was a then current definition of 'a concept', which governed its status in all knowledge including that of science. A concept was a mental (cognitive) representation (a way of calling to mind) of a set of named entities (several objects that have, or which are deemed to have, a real existence). It serves as a category (a way of classifying) that includes them, based on law-like regularities (analogies) between them. For example, the concept of 'force' can be used to explain how a wide variety of objects affect their behaviour. Medin and Rips (2005) then identify concepts, as a class, as having three broad functions. Firstly, they serve as the basis for the categorisation of knowledge: a wide range of individual entities is simplified by including them in a single concept, for this enables them to be readily identified. A concept thus enables any one of the series of entities that it includes to be more readily recalled from memory. Thus the concept of 'force' can be used to represent the collision of objects (for example, of snooker balls, of astronomical bodies, badly-driven vehicles). Secondly, they facilitate communication between individuals, provided that the same definitions of the concepts in question are accepted by the people concerned, so that the notion of 'the transmission of knowledge' then becomes possible. The understanding of a range of entities, including predictions of their behaviour, can thus more readily 'pass' between individuals when a single concept is used to represent all these entities. For example, using 'force' as the basis for sharing ideas about what happens in object-collisions. The use of concepts thus supports learning by an individual, as one person's concepts can be adopted by another person, because that second individual can base inferences on her/his existing and acquired concepts, and because existing concepts can be modified so as to incorporate new entities. For example, somebody who understands the game of snooker can readily understand the behaviour of entities on the Moon, for both can be explained with the use of Newton's Laws. Thirdly, concepts serve as the basis for memory, the efficient operation of which is considered to involve the establishment of meaningful relations between different concepts by what they call "semantic memory marriage" (Medin & Rips, 2005, p. 39). For example, the concepts of 'force', 'velocity', and 'acceleration', enable us to simultaneously understand the complex behaviour of objects when they move.

As a consequence of this view, concepts were seen and continue to be seen in science education as definite, law-like, and therefore bounded, mental representations that could readily be transmitted to others. This view of 'concept' was embedded in the use of behaviourist psychology to describe learning: a concept is treated as an object that can be 'given' to somebody else.

The Definition of 'Model'

Chapter 3 in this book was devoted to a discussion of views of the nature of models and of what creating them – modelling – involves. We noted that there are two broad schools of thought about what models are. Both are primarily concerned with their existence in the minds of individuals: 'mental models'.

In summary, the first and longest-standing view is what we call the 'representational' meaning, which also has its roots in behavioural psychology. In this meaning, mental models are depictions of objects, events, or processes that have a similar structure in some degree to what in the world-as-experienced (the 'real' world) is being represented. They are analogies of such entities and can therefore 'stand for' those entities. The second view we call the 'artefactual' meaning. In it, mental models are epistemic creations, human-made artefacts, usually materialised in some way for sharing with others, that attempt to depict the world-as-experience by imagining what it is like. In this book, we adopt this second meaning, for two reasons. Firstly, because it enables us to view a model as representing behavioural data but without knowing in detail anything about the nature of the entities that causes it. Secondly, because it enables us to use models as investigative tools, for they are partially dependent on both theories about and experience of 'reality'. This definition is congruent with the ideas embedded in that broad church of psychological theories known as 'constructivism'.

The Relation Between 'Concept' and 'Model' in Science Education

Ideas change over time. The ideas used in science education and derived from the natural sciences and the social sciences can be viewed as changing in one of two ways. At worst, they can be treated like fashions in clothing. That is, they are changed for no obvious functional reason, but perhaps because new research workers have not appreciated what has previously been done by others, or because some researchers have been more able to communicate their ideas than are others. At best, ideas evolve in a quasi-Darwinian manner so elegantly described by Thomas Kuhn (1996). Those concepts that after changes have taken place and which subsequently provide the more comprehensive and precise explanations of phenomena, will survive and be used ever-more extensively. Thus, if we arrive at definitions of 'concept' and 'model' in contemporary use that show a distinct difference between them, then both can continue to be used. However, if we cannot identify such a difference, then the more recently favoured word ('model') will supersede the longer-established word ('concept'). The issue then is the nature of an appropriate relationship between the two words.

The literature about 'concepts', 'concept formation', and 'conceptual change', in science education – which peaked in extent in the 1980s and 1990s – took a 'representative' view of 'concept' (as in the summary above). This was perhaps driven by the social imperative to identify precise meanings for individual words that to be readily communicated to students, a way of stemming the avalanche of individual facts that was submerging the school science curriculum, and a way of providing safe anchors for 'science for all' curricula. The 'transmission' approach to teaching such concepts, that implies the 'representative' meaning of concept, is still widely practiced despite the early evidence that it too-often fails (see, for example, Gilbert & Watts, 1983). The literature about 'models' in science education, which has grown steadily since the late 1990s (and which has perhaps not yet peaked) was initially dominated by an emphasis on 'key models' that were perhaps to be taught with 'representative' assumptions (Millar & Osborne, 1998). Since then, the 'artefactual' approach has gained ground, as supported both by the evidence underpinning our advocacy of a 'modelling-based teaching' approach and by an increased curricular emphasis on 'nature of science' (Australian Curriculum Assessment and Reporting Authority, 2010; Department for Education, 2014; National Research Council, 2012).

Successful learning does involve students acquiring some 'ownership' of what is being learnt, of seeing that the understanding achieved helps them to gain some mental control over the world-as-experienced (this issue was explored in Chap. 3). If this is accepted, then there seems every reason why the 'artefactual' view of models should be extended to that of concepts. But does this mean that 'models' and 'concepts' are the same thing, such that only one word ('model') should remain in use?

The 1980s work on 'concept' was bounded by its behaviourist definition, despite the gradual emergence of constructivist perspectives on learning. This behaviourist definition has underlain the 'concept understanding' research field until recently. It has focused on the use of single concepts to describe simple phenomena, those that can be defined precisely in terms of one simple and coherent set of law-like attributes, for example 'heat', 'light', 'force', 'work'. The gradual adoption of the artefactual meaning of concept enables a particular concept to have a more flexible meaning and to be used to actively interpret the world-as-experienced. This meaning would create the intellectual space within which individuals may explore their understandings and, one would hope, come to 'own' them if they prove useful in providing explanations.

The recent work on 'models' has focused on more complex phenomena, those that require several sets of bounded attributes (concepts) for their depiction, for example chemical reaction, evolution. In short, an artefactual model can be viewed as requiring the use of several artefactual concepts, for example a common model of chemical reaction can then invoke the concepts of substances, properties of matter, change. In short, by the adoption of 'artefactual' views, the meanings of 'concept' and 'model' converge. They can thus both continue to be used, the distinction only being in the complexity of the entity being discussed: simple entities invoke single 'concepts', complex entities invoke several 'concepts' within the framework of a 'model'.

If this is so, existing work on what is usually termed 'conceptual development-based teaching' can be updated so that is philosophically and pedagogically congruent with 'modelling-based teaching'. However, before doing so we must attempt to untangle the linguistic and cognitive web surrounding the processes by which concepts come into existence, evolve, and are changed, within an individual.

An Artefactual Perspective on Concept Formation, Evolution, and Change

Concept Formation and Evolution

The most intensive period of concept formation by humans covers the first few years of life. Most human babies – those blessed with the common attributes of mental functioning – make sense of their environment by finding and remembering and labelling regularities in their experience of objects, places, events, and people: they form concepts. These concepts change as a greater and more diverse range of experience is incorporated within them. This process of conceptual evolution is accelerated for young people when older people begin to repeat words in conjunction with the provision of or sharing of those experiences, for the words serve to bring diverse experiences together in memory. Different adults may use a word in different ways, thus providing a child with a range of meanings. As time passes, children are increasingly presented with novel, unexpected, experiences, closely associated with the words that link them together. Indeed, the words are often presented before, or in the absence of, the associated experiences. This is the basis for all pre-school, kindergarten, nursery, education, and indeed continues throughout the years of schooling. The concepts initially acquired (for example, 'book', 'place', 'house') clearly involve immediate perceptual judgement (for example 'short', 'near', 'big'), and may be called 'concrete concepts'. However, as time passes, the concepts being presented and acquired are derivatives of that experience, not obviously perceptually related to it, and may be termed 'abstract concepts' (for example 'acceleration', 'intensity', 'value').

As we have already said, the concepts acquired early in life evolve with the passing of time, not least because of being associated with an increasing diversity of experience that seems relevant to them. Inagaki and Hatano (2008) identify four ways in which initial concepts can change:

- A new concept emerges and subsumes the earlier version. For example, an earlier concept of 'force' based only on perceptions is replaced by one in which beliefs about the nature of causality play a part. For example, whilst seeing an object move can be described as it being forced, a later observation of the movement that has taken place can be said to have been caused by the operation of a force;

- A new concept emerges and supersedes the earlier version, although the latter continues to be held. For example, whilst a newer concept of 'force' is used in respect of object-movement, the older perceptual meaning (that is, what one can see happening) is remembered, if not actually used;
- A new concept emerges from the earlier version by differentiation, but both continue to be used because they relate to different domains of experience. Both meanings of 'force' co-exist after the domains in which it is experience separate. For example, encountering political force after years of experience of simple physical forces leads to a separation of these meanings;
- A new concept emerges by the integration to two earlier concepts. For example, one concept of 'force' becomes equally applicable to a wide variety of experiences and phenomena when the operation of a cause is seen to unify the two experiences.

The first of these alternatives is what the formal education system intends to support. The second is probably what occurs in practice during science education, however taught. The third alternative is very common during MBT when different groups of students produce different models (and hence their constituent concepts), whilst the fourth occurs at the end of the MBT sequence when the ideas of different groups have converged into a consensus model.

Although there seems to be no fixed sequence in which changes to initially formed concepts take place, what determines this process of evolution, natural or forced, can be illustrated for that range of experiences that fall within the domain of 'biology'. Inagaki and Hatano (2008) have suggested that change in initially formed biological concepts will involve the experience of several very diverse events: the acquisition of more experiential knowledge of biological systems; the acquisition of deliberately provided knowledge of the behaviour of such systems, such as is provided by a broadening of the types of biological phenomena experienced; the actually learning some of the abstract concepts used in biology; an increased awareness of how inference is based on the notion of conceptual hierarchies; the acquisition of a better understanding of what the operation of 'causality' entails. It will be noticed that some of these categories involve biological systems whilst others focus on the act of thinking as such and are therefore far wider in application.

Let us bring some of these ideas together. Those ideas that are entirely created from direct experience, by 'concept formation', may be tentatively called 'preconceptions', although the significance of the acquisition of external word-applications from adults is difficult to quantify. What happens to the concepts formed in the light of more extensive and diverse experience may be termed to be the result of the process of 'conceptual evolution'. This process of conceptual evolution is certainly accelerated by the deliberate and focused intervention of information from elsewhere (adults, teachers, media) such that 'conceptual change' may take place. Educational systems, as a consequence of the commonly-held assumptions about their purpose, currently concentrate on bringing about, or at least on trying to bring about, deliberate conceptual change of specific kinds: these are underpinned by a 'representation' view of 'concept'. Other individuals and agencies are seeking to promote the learning of 'concepts' as being 'artefactual' in nature, thus supporting their use in scientific creativity. 'Informal learning' resources, such as those provided in museums, zoos, botanical gardens, are particularly valuable in this respect (Gilbert & Afonso, 2015).

Conceptual Change

Studies of the understanding by high school students of a wide range of the concepts used in science reveal existing interpretations of them (misconceptions, alternative conceptions) which are associated with prior experience, in some way and to some degree, of the phenomena which they represent. We may therefore conclude that the process of changing those concepts does not generally involve either what Rusanen (2014) calls the *reinterpretation* (the use of analogies to other systems) or the *invention* (the creation of new concepts), of them to a much lesser extent than their *reorganisation,* a process in which the categories of knowledge into which they fall is altered.

The deliberate process of 'conceptual change' thus involves a reorganisation in which a series of conditions have to be met. Vosniadou, Baltas, and Vamvakoussi (2007) identify them as three in number. Firstly, the ontology that underpins the desired (that is, socially accepted) concept must be acquired, one that enables the relationships between its perceptual and physical properties to be understood in way that is precisely congruent with those of the established scientific knowledge that constitutes the school curriculum. For example, in doing so, learners would move from believing that matter is continuous to believing that it is based on the atomic model. Secondly, the epistemology that underpins the desired concept must be grasped. A commitment to the epistemology of naive realism ('things are just like that') has to be replaced by one what accepts that macroscopic phenomena are always the outcome of the action of entities at the sub-microscopic level. Thirdly, students have to become explicitly aware that they can mentally represent their knowledge of concepts and also be able to deliberately draw on the various 'modes of representation' to do so. How then can students be supported in making these altered commitments?

The Classical Approach to Bringing About Conceptual Change

The main approach to bringing about conceptual change that has been used in science classrooms was articulated by Posner, Strike, Hewson, and Gertzog (1982). It was concerned with addressing the existence of the 'misconceptions' that arise when the three conditions for conceptual change and hence learning (those of ontology, epistemology, and representation) are not fully met. Their model suggested that students be presented with (preferably) direct experience of a phenomenon and be required to make an interpretation of its behaviour. The model anticipated that those students who held misconceptions would be dissatisfied (or, through questioning, could be induced to become dissatisfied) with their explanations. This would open the way for another explanation, either provided by other students or the teacher, that seemed both plausible and more successful ('fruitful'), to be introduced. The 'acceptance' of the replacement – or modified – concept would then be consolidated by using it to make predictions of other properties of the same or allied systems that were subsequently found to be justified. The view that such 'acceptance' should be based on an active valuation of a concept, being thus artefactual in nature and not merely passive and being representative in nature, has steadily grown over the years since the 1980s.

In the 30 years since its introduction into classrooms – albeit a process that is still far from complete – a number of problems with the Posner et al.'s (1982) model have been identified (Vosniadou, 2012). Three of them stand out. Firstly, many individuals' grasp of and commitment to a concept has been found not to change immediately in the light of the experience provided, as had been anticipated. The process of change has been found to be partial, slow, and subject to intermittent reversal. The occurrence of the problem is probably exacerbated by the generally excessive content load of science curricula, which lead teachers to race through topics very rapidly ('Newton's Laws in 35 minutes'), a process which gives little time for changes in understanding and genuine valuation. Secondly, students all too often have a poor understanding of nature of science (see Chap. 9) and hence fail to see the epistemological significance of changing their ideas. Thirdly, there is clear evidence that conceptual change is not an entirely rational process, but one in which motivational and affective factors play major roles (Sinatra & Pintrich, 2003). Students have to feel that changing their ideas is something that will be of intrinsic value to them. Science educators must surely not accept the considerable failure of many – if not most – students to both understand and appropriately value the socially accepted scientific concepts. Whilst many of those concepts will remain permanently in scientific use, others of them will change. For the sake of what might be termed their 'conceptual health' students have to acquire the notion of the conditional and provisional value of the concepts that they learn. What, then, can be done to address these issues?

A Modelling Approach to Conceptual Evolution and Change

The reasons why attempts to support conceptual evolution and change currently are only very partially successful lie, we suggest, in the notion of 'concept' that underlies them. The 'representative' view of concept adopted seems to imply that science education involves students just learning the nature of the representations in which the subject is presented to them and being able to convince their teachers that they can do so. This may meet the 'ontological' and 'representational' requirements for conceptual change, but leaves the 'epistemological' aspects untouched.

The value of modelling in conceptual change in science education has been endorsed by Clement (2008), whilst its centrality in science itself (a precondition for acceptance in science education) was expounded by Nersessian (2008). But, in the context of the current discussion, two questions emerge: Would the adoption of the 'artefactual' view of concept and its explication through the 'modelling-based teaching' approach enable all the three conditions of conceptual change to be met? Would a modelling-based approach to conceptual change overcome the limitations to the 'classical model of conceptual change'? We suggest that if MBT is suitably taught, both questions can be affirmatively answered. The following ideas and examples support our view.

Ontological Condition

Independently of the approach adopted for models and modelling, research conducted in the last decade (like Barab, Hay, Barnett, & Keating, 2000; Buckley, 2000; Campbell, Zhang, & Neilson, 2011; Passmore & Stewart, 2002) has shown that students learn scientific consensus ideas when participating in MBT (from any of the perspectives discussed in Chap. 4). Studies in which MBT involve students in constructing a model de novo (like Khan, 2008; Maia & Justi, 2009; Mendonça & Justi, 2011; Núñez-Oviedo, Clement, & Rea-Ramirez, 2008) exhibit particular good results in terms of what we might term ontological learning. For instance, when secondary students who have never learnt about chemical equilibrium participated in activities designed and conducted from the Model of Modelling perspective (Maia & Justi, 2009), they developed a broad and consistent view on the topic. This allowed them: (i) to use the main ideas concerning chemical equilibrium (the coexistence of reactant and products, the dynamics of the process, the constant concentrations of reactants and products in the equilibrium state, and the alterations to equilibrium states that are possible) to discuss the behaviour of unknown systems in equilibrium; and (ii) to criticise ideas that expressed the main alternative conceptions on the topic that are discussed in the literature. We view both of these attainments as evidence that those students acquired the scientific ideas expressed in the school science curriculum.

Representational Condition

The abilities to represent ideas, to select the most suitable mode of representation to do so in a given context, and to translate a given model between the modes in which it can be depicted are at the core of 'visualisation'. They can naturally be focused on in MBT (as discussed in Chap. 7). This occurs particularly when the stage of *expressing the mental model* is recognised as essential in the process, as it is done when the Model of Modelling perspective is adopted. Studies focused specifically on issues related to this stage of modelling (Gilbert, Justi, & Queiroz, 2010; Justi, Gilbert, & Ferreira, 2009) show that:

- secondary students can acquire visualisation skills in respect of internal and external representations during MBT. For instance, when learning about ionic bonding, students were able to: understand the conventions of representations for all relevant modes (mainly the concrete and the symbolic ones), translate a given model between distinct models, construct representations within any relevant mode for a given purpose, and solve novel problems using representations previously built up or use versions derived from them. Additionally, evidence of students' expression of each of these skills were obtained not only in activities in which students were requested to express their models, but also in activities related to the other stages of modelling (Gilbert et al., 2010);
- specific teaching activities (for instance, the explicit request to use distinct modes of representation and to compare the scope and limitations of the models produced in each case) can support students' understanding of the role of representations in science, the idea that distinct representational levels of scientific knowledge exist, and specific ideas involved in some levels (mainly the sub-microscopic one). For instance, students were able to understand the dynamism of chemical equilibrium from the use of the simulations that they proposed based on the concrete models they had previously built. Those students were also able to understand that the main role of the sub-microscopic level is to explain the macroscopic level, and to use the representations that they had produced to do so (Justi et al., 2009);
- teachers' actions (for example, the questioning of students about the codes of representation used in their models) can allow for a comprehensive students' understanding of both the production of a given representation and particularities of the concept (or relationship) to be represented. For instance, when a group of students had their first ever experience of using coloured play dough to represent single molecules, they had not initially thought that colours might be used as a code of representation. In that context, they produced a concrete model for iodine molecules by using one red and one blue play dough piece. It was only from the teacher's questioning that they realised that I_2 means two bound atoms of iodine (Oliveira, Justi, & Mendonça, 2015). In fact, it was during the discussion of this code of representation that they have really understood the meaning of 'chemical element'.

Epistemological Condition

Epistemological aspects of conceptual change can be viewed as intrinsic to MBT, for the incorporation of education for and about nature of science in MBT (see Chap. 9), allied to the enhanced focus on visualisation (see Chap. 7) and argumentation (see Chap. 6) would inevitably lead to this taking place. In particular, students' views on models, modelling, and their role in the production of tentative scientific knowledge – students' meta-modelling knowledge – can be enhanced. The literature reports that students and teachers generally have naïve views on models (Carey & Smith, 1993; Grosslight, Unger, Jay, & Smith, 1991; Justi & Gilbert, 2002, 2003; Treagust, Chittleborough, & Mamiala, 2002). However, initiatives of involving students in MBT show a different picture.

Schwarz and White (2005) provided opportunities for students to create, evaluate, and revise computer models for physics topics (force and motion), and to engage in discussions concerning models and the process of modelling. Analyses of pre- and post-written tests and of post-instruction interviews with a sample of students showed that their understanding about the nature and purposes of models were broadened. On the other hand, there was no significant improvement in students' views on aspects more clearly related to modelling: model evaluation and revision. This result caused the authors to recognise that their pedagogical approach, their research instruments, or both of them, needed some modification. In the light of our current knowledge in the area, we suppose such unexpected results may have occurred due to the inherent limitations on computer modelling, since the expression, test, and modification of students' models are restricted by the constraints of the computer system. So, although explicit discussions about models and modelling had occurred, students may have had limited modelling experiences.

In a seven-months study in which students participated in three teaching sequences based on modelling (Maia, 2009), the development of their meta-modelling knowledge was analysed. Before participating in the modelling-based activities concerning chemical topics, students were involved in an activity in which they had to produce a model that explains how a hypothetical special glue works, one that can be used for all materials and that sticks them together instantaneously. At the end of the activity, students were also requested to describe how they had performed the task, that is, to describe the stages they had experienced when addressing the problem. As there was no specific scientific knowledge involved in this initial activity, and as it was performed at the beginning of the academic year, students could show their current, untutored, modelling knowledge and skills. During the MBT of the chemical topics, when all classes were video-recorded, sometimes the teacher asked students to justify their particular actions, and to compare their current actions with those they had described in the initial activity. Such answers always showed an increase in students' meta-modelling knowledge. More importantly, data were also collected by a series of three interviews with a sample of students (30 % of them) selected among those that had participated in the whole MBT process. The initial interview occurred immediately after the initial activity, the second one occurred after the MBT of the second chemical topic, and the last one occurred at the end of the academic year, a month after the end of all modelling activities. In all of them, students were questioned about: specific actions and/or answers during the modelling activities; the relationships they viewed between the process they were experiencing and the production of scientific knowledge; and their difficulties in performing the activities. Additionally, in the last session, each student was asked to model how a device in daily use (a microwave oven, or a spray can, or a digital blood pressure monitor, etc.) works, and to explain how s/he would convince someone else that her/his model was a good one. This task was included as an opportunity to express and use the meta-modelling knowledge and skills they had developed.

The integration of all this data supported the analysis of how students' meta-modelling knowledge increased from initial simple and naïve views (basically 'think about the phenomenon and then just produce the model') to complex and broad ones (those that included not only all the stages of modelling, but also the roles that both each of them and creativity played in the process). Finally, the fact that there was no explicit teaching of modelling *per se*, and the justifications provided by many students for their actions and ideas supported a conclusion that their participation in many activities in which they experienced all the stages of modelling, and the resulted learning, was determinant in the evolution of their meta-modelling knowledge.

Meeting Additional Conditions for Conceptual Change During MBT

As previously emphasised, the three main criticisms of the classical approach to conceptual change are related to the facts that: conceptual change does not occur overnight; students do not generally see the epistemological significance of changing their ideas; and conceptual change is an entirely rational process. We view MBT as a possible successful approach to overcome such limitations.

In terms of MBT approached from the Model of Modelling perspective, which is the one for which we have several sets of data from regular classrooms, the problems with the classical conceptual-change model are addressed. In these circumstances, there is:

- the realisation that students' learning is not an instantaneous and/or linear process. Our experience in planning, conducting, and analysing teaching situations from this perspective shows that students interpret evidence, establish relationships, visualise, etc. in different ways. This justifies, for instance, the impossibility of planning one single test hoping that it makes all students realise inconsistencies in their initial models. In the teaching sequence for ionic bonding, for instance, three tests are proposed and, sometimes, students themselves propose distinctly different ones. Data obtained when students perform such activities (for instance, those reported in Mendonça & Justi, 2011) show that the change from a 'NaCl molecule' (sic) to a 'NaCl lattice' model may be associated with the performance of any of them, and in an idiosyncratic way (that is, by the establishment of distinct relationships, and by the use of different arguments). In any case, the discussion of (i) the arguments that students produce to defend their models in the face of both what is presented in each test and their colleagues different ideas of how to interpret the same test, and (ii) their intermediate models (those that only partially incorporate new ideas or in which only secondary aspects of the model are changed), requires some time. The cognitive process involved in changing from one model to another requires not only the understanding of the scientific facts concerning how a given model can or cannot explain the properties of ionic substances, but also representational and epistemic issues that contribute to compose a network of knowledge from distinct natures (including meta-knowledge on the nature of models) (Vosniadou & Skopeliti, 2013);

- the recognition that models are always tentative. The development of students' meta-knowledge on models and modelling does include their clear appreciation that models (both scientific and their personal) are tentative, and that the coexistence of multiple models for a given entity in a given context is possible. However, this epistemological view is also not immediately acquired during the MBT sequence, and is definitely not acquired from the teachers' discourse. Rather it is developed from successive modelling experiences, as it was observed in two long-terms studies (Maia, 2009; Mendonça & Justi, 2013). In both of them, as far as students understood the tentative nature of models and the coexistence of multiple models, they performed the modelling activities more naturally, increasing their participation in the discussions – which also supported their learning;
- the recognition that affective issues are as important as rational ones in supporting students' learning. As in the case of students' enhanced epistemological views on models and modelling, the influence of affective issues also occurred gradually in the process, as a consequence of students having greater 'ownership' of their learning process. As time went by, students got themselves more involved in the process, which was reflected, for instance, in the increasing level of the arguments they expressed (Mendonça & Justi, 2014). This was evident, for instance, during the MBT for ionic bonding, when students expressed the most complex arguments (those that not only involve justifications supported by many evidence, but also aim at persuasion) only during the final activities, for here they felt really confident about their knowledge.

Concluding Remarks

We have argued that, if 'artefactual' views of both 'concept' and 'model', are adopted, then the meanings of the two words converge: for consistency of use in modelling-based contexts, only 'model' needs to be retained. If this is done, not only the models but also the concepts of which they are composed can be discussed in the same terms. By doing so, the precepts of MBT can be viewed as drawing on the constructivist traditions of teaching for concept development. For example, the testing of predictions made and the social processes involved in the production of concepts and of models are similar.

References

Australian Curriculum Assessment and Reporting Authority. (2010). *Science: Foundation to year 10 curriculum*. Canberra, Australia: Australian Curriculum, Assessment and Reporting Authority.

Barab, S. A., Hay, K. E., Barnett, M., & Keating, T. (2000). Virtual solar system project: Building understanding through model building. *Journal of Research in Science Teaching, 37*(7), 719–756.

Buckley, B. C. (2000). Interactive multimedia and model-based learning in biology. *International Journal of Science Education, 22*(9), 895–935.

Campbell, T., Zhang, D., & Neilson, D. (2011). Model based inquiry in the high school physics classroom: An exploratory study of implementation and outcomes. *Journal of Science Education and Technology, 20*(3), 258–269.

Carey, S., & Smith, C. (1993). On understanding the nature of scientific knowledge. *Educational Psychologist, 28*(3), 235–251.

Clement, J. J. (2008). The role of explanatory models in teaching for conceptual change. In S. Vosniadou (Ed.), *International handbook of research on conceptual change* (pp. 417–452). New York, NY/London, UK: Routledge.

Council, N. R. (2012). *A framework for K-12 science education: Practices, crosscutting concepts, and core ideas*. Washington, DC: The National Academies Press.

Department for Education. (2014). *The national curriculum in England - Framework document*. London, UK: Department for Education.

Gilbert, J. K., & Afonso, A. S. (2015). Lifelong learning: Approaches to increasing the understanding of chemistry by everybody. In J. García-Martínez & E. Serrano-Terregosa (Eds.), *Chemical education: Best practices, opportunities and trends* (pp. 123–148). Weinheim, Germany: Wiley.

Gilbert, J. K., & Watts, D. M. (1983). Conceptions, misconceptions, and alternative conceptions: Changing perspectives in science education. *Studies in Science Education, 10*(1), 61–98.

Gilbert, J. K., Justi, R., & Queiroz, A. S. (2010). The use of a Model of Modelling to develop visualization during the learning of ionic bonding. In M. F. Taşar & G. Çakmakcı (Eds.), *Contemporary science education research: International perspectives* (pp. 43–51). Ankara, Turkey: Pegem Akademi.

Grosslight, L., Unger, C., Jay, E., & Smith, C. L. (1991). Understanding models and their use in science: Conceptions of middle and high school students and experts. *Journal of Research in Science Teaching, 28*(9), 799–822.

Inagaki, K., & Hatano, G. (2008). Conceptual change in Naïve biology. In S. Vosniadou (Ed.), *International handbook of research on conceptual change* (pp. 240–263). New York, NY: Routledge.

Justi, R., & Gilbert, J. K. (2002). Modelling, teachers' views on the nature of modelling, implications for the education of modellers. *International Journal of Science Education, 24*(4), 369–387.

Justi, R., & Gilbert, J. K. (2003). Teachers' views on the nature of models. *International Journal of Science Education, 25*(11), 1369–1386.

Justi, R., Gilbert, J. K., & Ferreira, P. F. M. (2009). The application of a 'model of modelling' to illustrate the importance of metavisualization in respect of the three levels of representation. In J. K. Gilbert & D. F. Treagust (Eds.), *Multiple representations in chemical education* (pp. 285–307). Dordrecht, The Netherlands: Springer.

Khan, S. (2008). Co-construction and model evolution in chemistry. In J. J. Clement & M. A. Rea-Ramirez (Eds.), *Model based learning and instruction in science* (pp. 59–78). Dordrecht, The Netherlands: Springer.

Kuhn, T. S. (1996). *The structure of scientific revolutions* (3rd ed.). Chicago, IL: University of Chicago Press.

Maia, P. F. (2009). *Habilidades Investigativas no Ensino Fundamentado em Modelagem [Investigative skills in modelling-based teaching].* (PhD thesis), Universidade Federal de Minas Gerais, Belo Horizonte, Brazil.
Maia, P. F., & Justi, R. (2009). Learning of chemical equilibrium through modelling-based teaching. *International Journal of Science Education, 31*(5), 603–630.
Medin, D. L., & Rips, L. J. (2005). Concepts and categories: Memory, meaning, and metaphysics. In K. J. Holyoak & R. G. Morrisom (Eds.), *The Cambridge handbook of thinking and reasoning* (pp. 37–72). New York: Cambridge University Press.
Mendonça, P. C. C., & Justi, R. (2011). Contributions of the Model of Modelling diagram to the learning of ionic bonding: Analysis of a case study. *Research in Science Education, 41*(4), 479–503.
Mendonça, P. C. C., & Justi, R. (2013). The relationships between modelling and argumentation from the perspective of the Model of Modelling diagram. *International Journal of Science Education, 35*(14), 2007–2034.
Mendonça, P. C. C., & Justi, R. (2014). An Instrument for analyzing arguments produced in modeling-based chemistry lessons. *Journal of Research in Science Teaching, 51*(2), 192–218.
Millar, R., & Osborne, J. (1998). *Beyond 2000: Science education for the future.* London, UK: King's College, London School of Education.
Nersessian, N. J. (2008). Mental modeling in conceptual change. In S. Vosniadou (Ed.), *International handbook of research on conceptual change* (pp. 391–416). New York, NY: Routledge.
Núñez-Oviedo, M. C., Clement, J. J., & Rea-Ramirez, M. A. (2008). Developing complex mental models in biology through model evolution. In J. J. Clement & M. A. Rea-Ramirez (Eds.), *Model based learning and instruction in science* (pp. 173–193). Dordrecht, The Netherlands: Springer.
Oliveira, D. K. B. S., Justi, R., & Mendonça, P. C. C. (2015). The use of representations and argumentative and explanatory situations. *International Journal of Science Education, 37*(9), 1402–1435.
Passmore, C. M., & Stewart, J. (2002). A modeling approach to teaching evolutionary biology in high schools. *Journal of Research in Science Teaching, 39*(3), 185–204.
Posner, G. J., Strike, K. A., Hewson, P. W., & Gertzog, W. A. (1982). Accommodation of a scientific conception: Toward a theory of conceptual change. *Science Education, 66*, 211–227.
Rusanen, A.-M. (2014). Towards an explanation for conceptual change: A mechanistic alternative. *Science & Education, 23*(7), 1413–1425.
Schwarz, C. V., & White, B. Y. (2005). Metamodeling knowledge: Developing students' understanding of science modeling. *Cognition and Instruction, 23*(2), 165–205.
Sinatra, G., & Pintrich, R. (2003). *Intentional conceptual change.* Mahwah, N.J.: Erlbaum.
Treagust, D. F., Chittleborough, G., & Mamiala, T. L. (2002). Students' understanding of the role of scientific models in learning science. *International Journal of Science Education, 24*(4), 357–368.
Vosniadou, S. (2012). Reframing the classical approach to conceptual change: Preconceptions, misconceptions and synthetic models. In B. J. Fraser, K. G. Tobin, & C. J. McRobbie (Eds.), *Second international handbook of science education* (pp. 119–130). Dordrecht, The Netherlands: Springer.
Vosniadou, S., Baltas, A., & Vamvakoussi, X. (2007). *Reframing the conceptual change approach in learning and instruction.* Oxford, UK: Elsevier.
Vosniadou, S., & Skopeliti, I. (2013). Conceptual change from the framework theory side of the fence. *Science & Education, 23*(7), 1427–1445.

Chapter 6
The Role of Argumentation in Modelling-Based Teaching

Abstract The various stages in the modelling of a phenomenon involve the activity of argumentation. As the use of the Toulmin's model of argumentation has been found to present problems in the analysis of data from science education research, a simpler model was adopted for use in the context of MBT. An analysis of the skills and abilities used in the various stages of argumentation is presented. This is followed by an analysis of the role of argumentation in the various stages of modelling. This relationship between argumentation and modelling is then presented as a series of diagrams. The value of particular modes of representation in the conduct of argumentation during modelling is then discussed. Finally, the implications of these matters for teachers' support of modelling are discussed.

Argumentation

According to English dictionaries, the verb 'to argue' has three main meanings: to disagree with someone verbally, often in an angry way; to state, giving clear reasons, that something is true, should be done, etc.; and to persuade someone either to believe or not to believe in something, or to do or not to do something. In the scientific context, the meanings of argumentation as being related to justification and persuasion are important. Sometimes argumentation involves both of them simultaneously because claims require an increase in their credibility, which is done by an address to the reliability of the evidence that supports the justification. In short, argumentation is a key epistemic practice in science and hence in science education (Driver, Newton, & Osborne, 2000; Duschl, 2008b; Kuhn, 2010).

When thinking about the structure of an argument, undoubtedly the best known approach is that of Toulmin's Argument Pattern (TAP) (Toulmin, 1958), which was produced from a logical-philosophical perspective. According to it, an argument is an assertion together with its accompanying justification and it consists of six parts: data, claim, warrants, qualifiers, rebuttals, and backing. When argumentation started to become the focus of many studies in science education, some researchers have used TAP to analyse arguments produced by students and teachers (for instance, Bell & Linn, 2002; Kelly, Druker, & Chen, 1998). However, researchers soon realised that it was difficult to identify all the components in argumentative discourses, which motivated them to modify TAP in different ways (Chin & Osborne,

J.K. Gilbert, R. Justi, *Modelling-based Teaching in Science Education*, Models and Modeling in Science Education 9, DOI 10.1007/978-3-319-29039-3_6

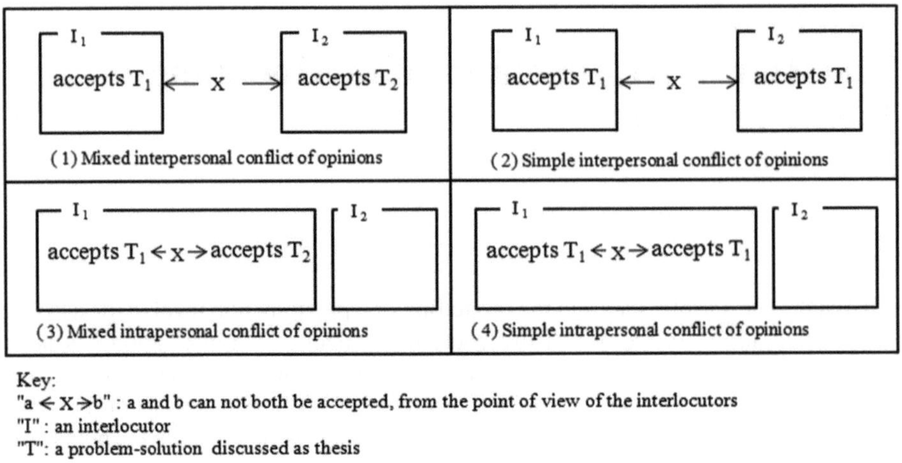

Fig. 6.1 Four possible argumentative situations (Baker, 2009, p. 131)

2010; Erduran, Simon, & Osborne, 2004; Zohar & Nemet, 2002). However, the analyses resulted from the use of such modified instruments continued showing some limitations. These were in terms of supporting discussions concerning the meaning and content of students' arguments into specific contexts, and/or their quality in terms of expressing ideas for specific purposes, including showing students' epistemological understanding (Duschl, 2008a; Mendonça & Justi, 2014; Sampson & Clarke, 2008); as well as concerning discursive situations occurred in contexts of where the generation of new knowledge takes place (Jiménez-Aleixandre & Erduran, 2008). In order to support such discussions, it seems more useful to adopt a simpler structure for arguments, according to which they are composed by a claim and its justification, the latter being supported by evidence (Kuhn, 1991). Therefore, the content of each of these elements, as well as the relationships between them and how they are used in a dialogue (which is influenced by the individual's skills of arguing, producing alternative theories, counter-arguing, and refuting) become significant in the characterisation of one's argumentation.

According to Jiménez-Aleixandre and Erduran (2008), in the science education context, another relevant dimension of argumentation is that of dialogue, since classrooms are communities where discussions are (or should be) promoted, where students' and the teacher's ideas can be produced, communicated and analysed. Baker (2009) identified four types of contexts in the discussion of ideas. These involve the mixed and simple interpersonal conflict of opinions, and the mixed and simple intrapersonal conflict of opinions (see Fig. 6.1).

Type 1 occurs when two individuals have distinct and different opinions. When arguing, each of them tries to justify the plausibility of his/her ideas, as well as pointing to the limitations of the other person's ideas. It is expected that, during the discussion, they reach a consensus. Type 2 occurs when one individual defends an idea, and the other person, despite not having his/her own idea, does not agree with the former. Therefore, in this argumentative situation, one individual tries to support his/her idea, whilst the other tries to identify flaws in it until they reach an agreement. Type 3 is characterised as a situation in which an individual believes in two possible divergent ideas and, from a discussion with someone who does not have an opinion about the subject, tries to reach a personal consensus about each of the ideas is better in a given context. Finally, Type 4 occurs when someone is not sure about the credibility of an idea and tries to reach a conclusion from discussion with someone who does not have an opinion about the subject.

In our view, intrapersonal contexts (Types 3 and 4 in Fig. 6.1) can also involve just one person. This means that the individual can have one or more ideas and, being in doubt about them, analyses them (by processes of comparing and contrasting) in order to decide which one should be used. This view is supported by Kuhn's (1991) idea that argumentation only involving one person has a 'social' component because the individual seeks to predict someone else's possible opinions during the analysis of his/her own ideas. In other words, one can either engage in a personal dialogue with possible solutions that s/he provides to a problem or one can predict distinct solutions that may be proposed by other people. In the normal context of schooling, where students work on their own, Type 4 contexts must be very common.

Argumentative Skills

In the introduction of her influential book, Deanna Kuhn asserts that:

> The ability not just to think, but to think well is, or at the very least should be, essential to fulfilled adult life. (…) Although much has been learned, there is still a good deal of human thinking about which we know almost nothing. (Kuhn, 1991, p. 1)

This motivated her to develop a broad study (reported in her book) about the skills used by people when thinking in ordinary situations. In this study, she assumed that argumentation is at the core of thinking. On account of this, she identified some skills as inherent in the act of arguing: *to deal with evidence*, that is, to identify, interpret and use it; *to argue*; *to counter-argue*; *to produce alternative theories*; and *to refute*.

Mendonça and Justi (2009) assume that each of the skills identified by Kuhn involves a series of abilities. Therefore, each of them was characterised as following.

To Deal with Evidence Following Bravo, Puig, and Jiménez-Aleixandre (2009), evidence is understood as observations, facts, signs, or data with which we can try to support an opinion. So, the skill of dealing with evidence is related to:

- meta-knowledge about the nature of evidence, that is, to know the meaning of 'evidence', and to identify both its main functions, and the criteria that can be used to assess the reliability, the specificity, and the sufficiency of evidence;

- the abilities of: identifying observations, facts, signs or data that can be used as evidence in a given context; planning investigative experiments that can produce evidence related to a given problem; selecting the most relevant evidence (that is, that which is specific, sufficient, and reliable) for a given purpose (Jiménez-Aleixandre, 2010); articulating distinct evidence with a justification in order to produce a coherent argument.

To Produce Arguments As previously mentioned, in the scientific context, argumentation can be viewed as a social process of justifications of claims, which are made in respect of data and of scientific theories. This process has the assessment of knowledge as its key aspect. Thus, scientific argumentation is directly related to justification and persuasion (Jiménez-Aleixandre & Erduran, 2008). Justification implies both the selection of evidence from available (empirical or theoretical) data and the adequate use of it in justifying knowledge. On the other hand, persuasion requires a rhetorical use of speech or writing with the aim of (i) balancing one justification against another; and (ii) strengthening claims in order to convince someone else, that is, with the aim of emphasising why a given piece of knowledge is better than others in a given context, or why it has greater explicative power. Therefore, a scientific argument can be defined as a claim properly supported by empirical or theoretical justification(s), and its strength will depend on the number and quality of the relationships established between the claim and its justification(s) (Mendonça & Justi, 2014). As a consequence, the production of an argument in the scientific context requires the abilities of: identifying evidence; differentiating data and/or evidence from causes; thinking critically about evidence; making justifications from clear relationships between (a piece of) evidence and a claim; drawing analogies that may be used as (parts of) justifications; providing the basis of justifications; using proper (verbal or otherwise) scientific language in the expression of the argument.

To Counter-Argue According to Kuhn (1991), to counter-argue means to imagine conditions that may falsify an idea that one defends. This occurs because we not only use evidence to support our ideas, but we can also identify new evidence that contradicts our ideas or even use already available evidence in a way that contradicts such ideas. In this sense, the skill of counter-arguing plays a crucial role in science because when we try to predict (or identify) evidence that may be contrary to our own ideas, it is not likely that we defend naïve ideas. This may also reflect a less naïve view of science, since to counter-argue implies a commitment to a view that knowledge is tentative. Additionally, if we can predict or identify evidence that is contrary to our own ideas (or that can decrease their strength when they were used in a distinct context), other people may also be able to do it. So, if we are able to counter-argue, it may be easier for us to defend our arguments in a discussion with someone else. Therefore, the skill of counter-arguing involves, besides all the skills required to produce an argument, those of: understanding how an argument can be challenged; analysing evidence from other points of view; identifying limitations of the evidence and/or analogies used in the justification of an argument; identifying the limitations of the relationships established between such evidence and/or analogies and the original claim; identifying the scope of the original argument.

To Produce Alternative Theories This skill involves one imagining an idea that is distinct from the one s/he believed in initially and that is equally plausible. To predict possible alternative theories is an important skill in argumentative practice because on many occasions a given outcome can result from a series of causes. In science, there are a lot of cases of the coexistence of alternative theories in some historical contexts and, after the passage of time, some of them resulted in meaningful advances in scientific knowledge. On the other hand, in ordinary life, to be a critical citizen involves being able to consider that a socio-scientific problem can have (and often has) more than one answer. So, the production of alternative theories is relevant in all argumentative contexts. In addition to the skills and abilities listed above, the skill of producing alternative theories involves the ability of accepting two or more explanations for, or points of view about, a given situation, by using one or another depending on their adequacy to the context.

To Refute Very often, argumentative practice involves the discussion between two individuals who defend different opinions about a given question with the aim of reaching a consensus on which opinion is the most adequate in a given context. The ability to defend one's idea as the most adequate to solve a given problem, that is, of refuting an idea that is alternative to one's own idea, is essential to argumentative practice. Two main reasons can be emphasised to show the importance of refutation:

- It is essential in the practice of citizenship, since many of the decisions that we make in our ordinary life involve choosing between two (or more) possible options.
- The existence of controversies between two individuals, or groups of individuals, who defend theories that are alternative to each other is essential to the building of scientific knowledge.

According to Kuhn (1991), there are two ways an individual may refute: (i) by producing arguments that are contrary to the idea of the second individual (those that are alternative to his/her own idea); or (ii) by simultaneously producing arguments that show how adequate is his/her idea and how inadequate is the second individual's idea. In the studies conducted in our Brazilian research group, we also consider that one can refute when expressing new arguments that support her/his original idea because we view this practice as an attempt to show that her/his original theory is more adequate than the alternative theory.

In terms of abilities for the practice of refutation, in addition to those mentioned in previous skills, we think it is essential that the individual could be willing to listen to someone else; to be able to critically analyse two distinct arguments without emotional interferences; and to be able to use language resources in order to emphasise aspects of a given argument.

Table 6.1 Argumentative skills and abilities related to them

Skills	Abilities
To deal with evidence	1. To identify facts, observations, signs or data that can be used as evidence in a given context.
	2. To plan investigative experiments that may produce evidence related to a given problem.
	3. To select the most relevant evidence (that is, those that are specific, adequate, and reliable) for a given purpose.
	4. To clearly interrelate distinct data to produce evidence.
To argue	(1–4) +
	5. To differentiate data and/or evidence from causes.
	6. To critically reflect on evidence.
	7. To propose justification from clear relationships between the evidence and the claim.
	8. To draw analogies that can be used as (parts of) justifications.
	9. To support the justifications.
	10. To use proper (verbal or not) scientific language in the expression of the argument.
To counter-argue	(1–10) +
	11. To understand how an argument can be challenged.
	12. To analyse evidence from other points of view.
	13. To identify limitations of the evidence and/or analogies used in the justification of an argument or of the relationships established between such evidence and/or analogies and the original claim.
	14. To identify the scope of the original argument.
To produce alternative theories	(1–14) +
	15. To accept two or more explanations for, or points of view about, a given situation, by using one or another depending on their adequacy to the context.
To refute	(1–15) +
	16. To listen someone else.
	17. To critically analyse two distinct arguments without emotional interference.
	18. To use language resources in order to emphasise aspects of a given argument.

Justi, 2015, p. 38

In order to facilitate future references to the skills and abilities required for an individual participation in argumentation, they are summarised (and numbered) in Table 6.1.

The occurrence of argumentation as shown in the use of the skills and abilities listed in Table 6.1 is consistent with a view of science as a human endeavour (i) motivated by the curiosity of understanding the world; (ii) the main outcome of which is the production of knowledge that may change in a dynamic and non-linear way in distinct historical, social, political, and economic contexts. This is justified by considering that, from such a perspective, (i) it is intrinsically necessary to justify and to judge the adequacy of models and theories in accordance with the available knowledge and evidence; (ii) science has a strong social nature, that is, the constructs produced by scientists are open to discussion and refutation by the scientific community (Mendonça & Justi, 2013a).

Argumentation in Science Education

Research on argumentation in science education has been mainly conducted in the last two decades. In the introduction to a book that synthesises classroom-based research in this area, Jiménez-Aleixandre and Erduran (2008) emphasise that the introduction of argumentative situations in regular lessons can contribute to support students'

> "access to the cognitive and metacognitive processes characterising expert performance (...); development of communicative competences and particularly critical thinking (...); achievement of scientific literacy (...); enculturation into the practices of the scientific culture (...); development of epistemic criteria for knowledge evaluation (...); development of reasoning based on rational criteria (Jiménez-Aleixandre & Erduran, 2008, p. 5).

Among with a series of other relevant conclusions, studies that analyse argumentation in science education have shown that the engagement of students in argumentative situations can also contribute to their conceptual development (for instance, Allchin, 2011; Chin & Osborne, 2010; Cross, Taasoobshirazi, Hendricks, & Hickey, 2008; Irwin, 2000). According to these authors, students' engagement in argumentation can promote a clearer understanding of the concepts already existing in their cognitive structure, so that new ideas can be integrated into the previous ones in order to enhance their knowledge. This can be explained by considering: the dialogical nature of the social production of knowledge; the existence of situations that involve reflection and assessment of knowledge inherent in argumentative teaching contexts (Hogan, 2000); the fact that argumentative situations support students' understanding of how we know what we know (Duschl & Osborne, 2002) and why a given knowledge is accepted in the scientific community. In other words, students' engagement in argumentative situations can support their learning about science (as clearly shown in other publications like Deng, Chen, Tsai, & Chai, 2011; McDonald & McRobbie, 2012). This occurs particularly when the argumentative situations involve the construction of explanations and the evaluation of evidence – which, in a class, can result from the interpersonal or intrapersonal conflict of opinions.

Finally, Jiménez-Aleixandre and Pereiro Muñoz (2002) show that argumentation may provide opportunities for empowering students' capacity to make decisions, so that they could act as a knowledge-producing community, rather than as knowledge consumers. This would happen if argumentative situations involve students in:

- generating hypotheses, conclusions, solutions and questions when solving problems;
- analysing and choosing between two or more possible explanations;
- analysing empirical evidence in the light of their previous knowledge;

- using theoretical or empirical evidence to support their claims;
- talking and writing about science when discussing how they solve problems, formulating hypotheses, expressing their ideas, and trying to persuade others that their ideas are reliable and useful in a given context (Deng et al., 2011).

Relationships Between Argumentation and Modelling

Independently of the characterisation of modelling, there is no doubt that argumentation is inherently part of it. From the stages of producing a model to the ones of accepting and using it, several aspects support (social or individual) discussions about the adequacy of (distinct parts of) a model. Thus, MBT contexts facilitate the occurrence of several argumentative situations involving a single student (intrapersonal context), a group of students or students and their teacher (interpersonal contexts). By simultaneously taking into account the identification of the argumentative skills and the abilities related to them (as discussed previously in this chapter), and the discussion on the stages of modelling from distinct perspectives (Chap. 3), we now discuss and exemplify the relationships between them in specific MBT contexts.

Relationships from the Current Literature

Most researchers (for instance, Böttcher & Meisert, 2011; Jiménez-Aleixandre, 2010; Windschitl, Thompson, & Braaten, 2008) recognise the occurrence of argumentative situations during the stages of testing and modification of models. This is so because it is during such stages that the explicative powers of distinct models compete to explain a given aspect, or to solve a given problem. As clearly emphasised by Böttcher and Meisert,

> Argumentation is a process of critical evaluation of models in the sense of verifying the appropriateness of one or multiple rival models according to their logical coherence and the available, empirical data. (Böttcher & Meisert, 2011, p. 111)

On the other hand, two studies show the occurrence of argumentative situations in other stages of the modelling process. Passmore and Svoboda (2012) stress that, in the scientific context, to model is an inherently argumentative action since

> … in virtually all aspects of modelling, from developing a question to judging between competing models that might answer that question, an individual is engaged in persuasive acts (…) They may be mental, written or oral, but they are about judging ideas and making sense of them; convincing oneself or others that the ideas and ways of looking at and explaining a phenomenon are useful. (Passmore & Svoboda, 2012, p. 1535)

Passmore and Svoboda claim that argumentation is part of such epistemic processes when considering the key role of models in asking questions, recognising data patterns, producing explanations for data, and proposing criteria for judging knowledge claims. For them, this is justified because such processes involve one or more of the goals for constructing scientific explanations and participating in argumentative situations that were proposed by Berland and Reiser (2009): *'sensemaking'*, which occurs when the individual tries to establish concise relationships between evidence and claims; *'articulating'*, which is related to the individual's understanding of explanations concerning how or why something occurs; and 'persuading', which occurs when the individual tries to convince others about the scientific accuracy and explicative power of his/her explanations. The connection between the epistemic practices involved in modelling and the three goals can also be established because students can achieve such goals when they interpret phenomena, search for evidence related to the aspects studied, propose relationships between data and models, and evaluate models – actions that are performed during the modelling stages that the authors name the *construction*, *use*, *evaluation* and *revision* of models.

The other study that relates argumentation with all the stages of modelling was undertaken in our Brazilian research group. It involved students in participating in MBT for ionic bonding and inter-atomic interactions (Mendonça & Justi, 2013b). In this study, the teaching activities were designed and conducted from the Model of Modelling Diagram (Fig. 3.2), and followed the characteristics of MBT as based on this framework (as discussed in Chap. 4). This means that students had the opportunity to experience all the stages we considered essential in modelling: *production*, *expression*, *test*, and *evaluation* of models. The original aims of this study were to investigate: (i) the quality of students' arguments when involved in such a MBT context; (ii) possible relationships between their arguments and the stages of modelling in relation to which they were expressed, and (iii) if and how the participation in modelling activities influence students' arguments.

In terms of the second aim (which is more directly concerned with the topic focused on in this chapter), Mendonça and Justi's study shows the occurrence of argumentative situations when students performed all the stages of modelling. Such situations particularly emphasise how students articulate different pieces of knowledge when producing their models; how they fit their mental models into an expressed model, and vice-versa; how they justified their models in terms of the previously defined aims or the questions posed in the tests undergone by the models; how their models were tested in the light of new (empirical or theoretical) data; how the scope and limitations of their models were evaluated (Mendonça & Justi, 2013b).

In sum, although the authors of these two studies view modelling from different perspectives, both of them identified the occurrence of argumentation when students engage in all stages they assume to be involved in modelling. However, the relationships between argumentation and modelling established in both papers are general ones. This means that the current literature does not present discussion of the more detailed and complex aspects, like the argumentative skills needed to underpin students' performance of actions involved in modelling.

New Relationships Involving Argumentative and Modelling Skills

A new analysis of the data gathered in the study discussed in Mendonça and Justi (2013b) allows us to identify and discuss the argumentative skills required for the performance of each of the modelling stages. In order to support such a discussion, we consider the skills and abilities required for the performance of each of the modelling stages (discussed in Chap. 4 and synthesised in Table 4.1).

Stage of Creating the Proto-Model

According to Table 4.1, the creation of the proto-models requires the student(s) to display the following skills and abilities:

A. To define the aims of the model or to understand the aims proposed by someone else.
B. To search for information about the entity to be modelled (in the modeller's cognitive structure and/or from external sources: bibliography, empirical activities, etc.).
C. To select and organise the available information by having the aims of the model in mind.
D. To use analogical reasoning or a mathematical tool on which to base the proto-model.
E. To integrate all the previous information and experiences in the creation of a proto-model.

In Mendonça and Justi's (2013b) study, when secondary students participated in MBT for intermolecular interactions, they have the opportunity to observe the distinct behaviour of two substances (iodine and graphite) in which the atoms are linked by the same type of bonding. When they were asked to observe and compare the behaviour of the two substances when heated, they obtained important information about the entities to be modelled, and could understand the purpose of the models to be created: to explain the distinct behaviour of the two substances before and after heating. In other words, these initial actions entailed sense making (as defined by Berland & Reiser, 2009).

Thus, the stage of creating a proto-model requires that students use the following argumentative skills:

- *to deal with evidence*, either from the empirical domain (obtained from data collected in the activity or asked for by the students – like the value of the melting temperature of the substances) or theoretical ideas (their previous ideas about melting temperature and the particulate model of matter);
- *to argue*, by producing and using arguments that clearly related different pieces of evidence (for instance, by establishing relationships between empirical evi-

dence and the kinds of bonds existing in the substances), and using that to provide justifications for their models.

These skills involve abilities 1–10 in Table 6.1.

Students' use of these skills is shown in many parts of the transcriptions of the groups of students' discussions, for example:

"In iodine, the bonds will be broken because its melting temperature is low. In the case of graphite, the particles only vibrate because its melting point is high." (Creation of a proto-model and its expression, stages in which students selected evidence that they judged as relevant, critically analysed the meaning of such evidence, and coherently articulated them with their previous knowledge. This resulted in the production of an argument that was expressed in clear language.)

"In the iodine, the bond is covalent because each atom has seven electrons." (Justification for the type of bonding represented in a model, a situation in which students based it on their previous knowledge.)

Stage of Expressing the Proto-Model

As stated in Table 4.1, the performance of this stage requires the students to display the following modelling skills:

F. To visualise the proto-model.
G. To use any of the models of representation (or a combination of them) to express the proto-model.
H. To adapt the expressed model to the mental proto-model (or vice-versa) by making comparisons between each other and/or between them and the aims of the model.

By considering both the discussions that generally occur when such modelling skills are used, and the fact that, in terms of the Berland and Reiser's (2009) goals for constructing scientific explanations and participating in argumentative situations, the main focus of this stage is 'articulating', the involvement in this modelling stage also requires the following argumentative skills:

- *to argue*, in the context of justifying the modes and specific codes of representation used in the expression of the model;
- *to produce alternative theories* that, in this context, would be distinct models for the same entity.

Due to the characteristics of the arguments produced in this stage, the use of such skills would mainly require the abilities 8, 9, 10, and 15 (Table 6.1).

In Mendonça and Justi's (2013b) data, the use of these abilities was observed, for instance, when the students:

- justified the use of different materials (different sized polystyrene balls, different coloured play dough) to represent distinct atoms and molecules;

- expressed the same mental model using several modes of representation (concrete, two-dimensional (drawings), and verbal);
- adequately used each of these expressed models for distinct purposes (that is, to explain distinct aspects or to emphasise specific details in their representations);
- clearly expressed analogical reasoning involving a symbolic representation of a model *("If iodine is represented by I_2, graphite is represented by C_2.")*

Stage of Testing the Model

According to Table 4.1, the performance of this stage requires the following skills:

I. To plan and conduct thought experiments.
J. To plan and conduct empirical experiments.
K. To critically analyse the results of the tests.
L. To modify or to reject the model.

When one is working individually, these skills are related to the Berland and Reiser's (2009) goals of 'sense making' and 'articulating'. When the modelling task is being performed in groups, they can also involve 'persuading' colleagues from the same group.

After the creation of the initial model to explain the distinct behaviour of iodine and graphite, students had the opportunity to conduct thought and or empirical experiments in order to confront their models with new empirical evidence obtained from the reaction between iodine and starch. Depending on how they interpret the results of such tests, their models were changed or rejected.

In order to perform the tests, students used the skills of:

- *dealing with evidence*, mainly by identifying the new empirical evidence and interpreting them in the light of their previous knowledge;
- *arguing* in favour of their new models when communicating and discussing them with the whole class;
- *counter-arguing,* by producing arguments different from the original ones and then proposing thought experiments that could be used to test their models;
- *producing alternative theories*, mainly in groups discussion, when students were trying to change their models for any reason.

The use of these skills implies in the use of abilities 1–15 (Table 6.1).

Stage of Evaluating the Model

The performance of this stage requires the following skills (Table 4.1):

M. To identify the limitations of the model by trying to use it in different contexts.
N. To identify the scope of the model by trying to use it in different contexts.
O. To convince others of the validity of the model.
P. To persuade others to use a given model.

The complexity of this stage requires the use of all the skills of argumentation, with special emphasis on 'to refute' aiming at persuading others. Such argumentative skills can be used in order to achieve all the goals proposed by Berland and Reiser (2009) for participating in the argumentative situations of 'sense making', 'articulating', and 'persuading'.

Following the results of the tests and discussions around them, students were convinced about the validity of their models. Then, they were able to produce well-justified arguments to emphasise the scope of their models. In the case of the models for the ionic bond, for instance, this was done in distinct ways, that is:

- by presenting arguments that were contrary to a given model that had been previously produced by the group:
  "The 'NaCl molecule' model does not explain the melting temperature because it does not show all particles close to each other and strongly attracted."
- by simultaneously expressing arguments that show the adequacy of the group's model and the inadequacy of other models:
  "The lattice model explains why the particles are interacting and are close to each other. The 'NaCl molecule' model is not consistent because if the interactions between the particles were equal, all of them should be bound."
- by expressing arguments that emphasise the qualities of the group's model and/or (implicitly or explicitly) emphasising that only that model has such characteristics:
  "Our model is plausible because it shows particles close to each other, with several attractive forces which strength varies with their distance."
  "Only the lattice model explains the value of the energy because in that model the particles are bound and organised."

Although some of these arguments, like some others quoted in this sub-section, contain ideas and/or were expressed using language that is not scientific acceptable, they were expressed during the learning of the chemical content. Therefore, this does not decrease the validity of their arguments at that moment.

Finally, as emphasised in Chap. 4, MBT can also contribute to foster students' participation in a quasi-authentic scientific practice by requiring them to work in small groups. Thus, all arguments are expressed by students in the contexts of group discussions or discussions between the teacher and the students – in other words, in social contexts. This implies that MBT provides opportunities for students to experience simple and mixed interpersonal conflict of opinions (In Baker's (2009) terms). On the other hand, as reasoning is an individual process, that is, as each student thinks of and establishes idiosyncratic relationships between distinct ideas (whether generated by the individual or expressed by someone else), it is possible

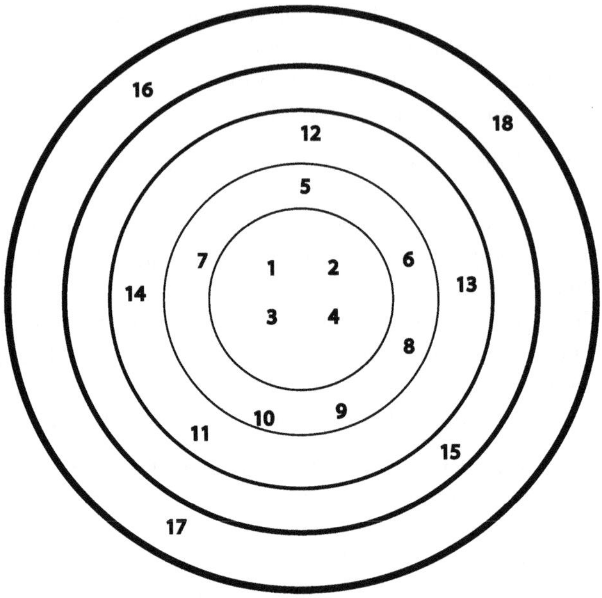

Fig. 6.2 Representation of the abilities involved in each of the argumentative skills

that in all the stages of modelling, but mainly during the production of the proto-model, and the planning and/or performance of thought experiments, each student experiences intrapersonal conflicts of opinions. However, contrary to the case of interpersonal situations, which can be clearly identified and characterised from data collected in class discussions, it is hard to have access to evidence of the occurrence of intrapersonal situations. This would be possible only if, for any reason, a student spontaneously expressed his/her line of reasoning at a given moment, or if the teacher (or a researcher) would ask the student to explain his/her thoughts and the student clearly verbalised his/her personal internal discussion. Therefore, although recognising that argumentative situations permeate the whole modelling process, it is not possible to associate the performance of specific modelling stages, thus with the use of given modelling and argumentative skills, with the types of argumentative situations proposed by Baker (2009).

Graphical Representation of the Relationships Between Modelling and Argumentative Skills

In order to facilitate the reader visualization of the relationships established in the previous sub-section, we designed a simple graphical representation. Additionally, in order to support the understanding of all its codes of representation, we opted for expressing and explaining it in parts (Figs. 6.2 to 6.5).

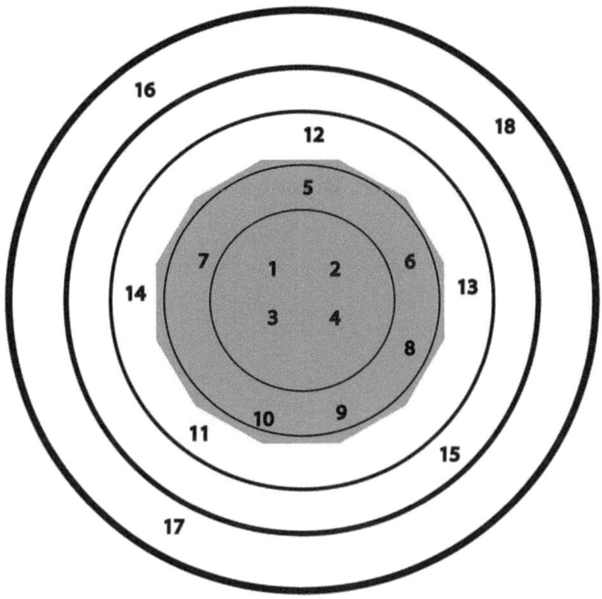

Fig. 6.3 Representation of the skills and abilities required for the creation of the proto-model

In this part of the chapter, the relationships were considered in the opposite order to that in which they were presented in the previous sub-section, that is, we start from the argumentative skills, represented by concentric and different colour circles (Fig. 6.2) rather than from the modelling stages. In this figure, the numbers are those identified in Table 6.1. So, they represent the abilities involved in the performance of each of the skills. The circles have different weights just to emphasise that they are associated with distinct skills.

Next, we represent each of the stages of modelling by using grey-scale geometric shapes. Again, distinct tons were used just to emphasise that the geometric shapes are related to specific modelling stages. Then, we superimpose such shapes onto the concentric circles that represent the argumentative skills (or parts of them) in order to show which abilities should be used by an individual when experiencing modelling (Figs. 6.3, 6.4, 6.5, and 6.6).

Figure 6.6 shows all the relationships established in the previous sub-section. It illustrates how the use of argumentative abilities and skills contribute to the performance of specific stages of modelling or, in the reverse order, how the participation in modelling activities requires the use of the argumentative skills and abilities. Emphasis is also placed on the facts that the argumentative situations that occur during each of the modelling stages can involve (i) interpersonal or intrapersonal contexts (as defined by Baker, 2009 and synthesised in Fig. 6.1), and (ii) discussion about two or more distinct models, or even a single model. Additionally, some of the intrapersonal contexts (mainly those related to both the creation of the proto-model and the design and conduct of thought experiments) can involve a single

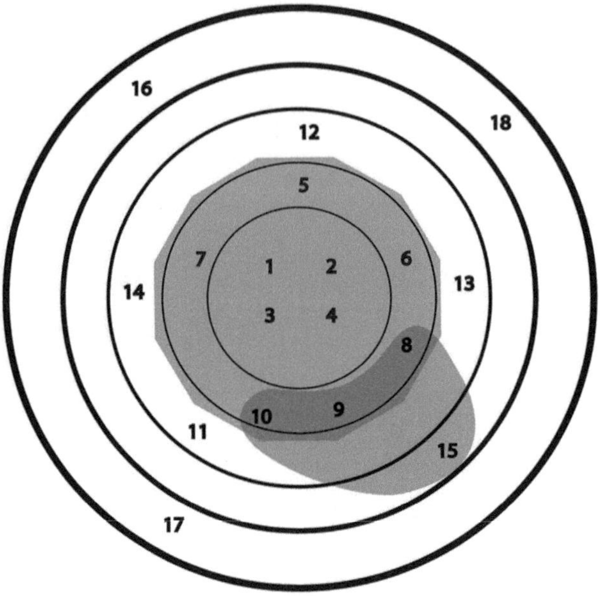

Fig. 6.4 Representation of the skills and abilities required for the creation and expression of the proto-model

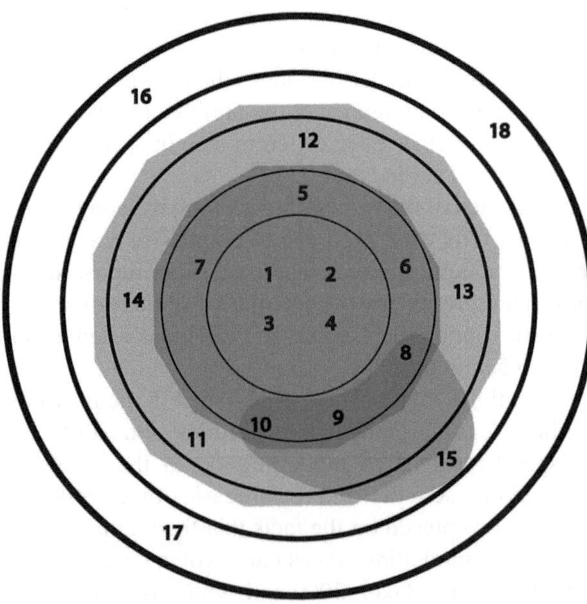

Fig. 6.5 Representation of the skills and abilities required for the creation and expression of the proto-model, and test to the model

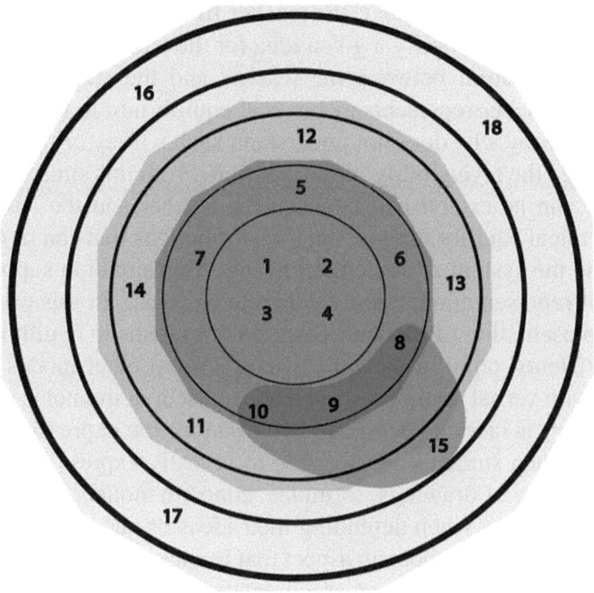

Fig. 6.6 Representation of the skills and abilities required for the creation and expression of the proto-model, test to the model, and the evaluation of the final model

student. In such cases, the student had to use not only the skills related to the particular stage, but also clearly distinguish his/her own arguments in favour and against a given idea or model.

New Relationships Between Argumentation and Modelling Involving Modes of Representation

Scientific discourse is multimodal. Scientists produce, discuss, and communicate their work by using (oral or written) texts, graphics, diagrams, three-dimensional models, mathematic equations, etc. This means that scientific arguments can also be expressed through modes of representations distinct from the verbal one (Cambrosio, Jacobi, & Keating, 2006; Lemke, 1998). Despite this, the absolute majority of studies in scientific argumentation that have been published focus on verbal argumentation.

As discussed in Chap. 3, a range of modes of representation can be involved in modelling. In some cases, the use of such modes was essential for the development of scientific knowledge. Watson and Crick's use of concrete models is a typical example. Thus, several questions arise: How does the use of particular modes of representation influence the conduct of argumentation? Are the relationships between scientific modelling and argumentation affected by the expression of arguments in a (partially at least) non-verbal mode?

In the science education context, the teacher frequently turns to representations in order to explain or to simplify a given idea for the students. When this generates discussion, argumentation between the teacher and the students can occur. For instance, when a teacher represents a chemical equilibrium in the form of an equation, a student can say s/he does not understand such a transformation because it is not possible to see the reversibility of the reaction. From this situation, an argumentative dialogue can be established between the teacher and the student about the concept of chemical equilibrium and the transformations that can or cannot be seen when observing the system at particular instants. This situation supports the use of other modes of representation, mainly concrete or visual. In this case, these could support the representation of particular aspects of systems in equilibrium.

From the students' point of view, the use of several other modes of representations together with verbal argumentation tends to occur at moments when their conceptual vocabulary is incomplete or inappropriate for the expression of their ideas. In other words, when students are not able to verbally express their ideas about a given subject, the use of drawings, formulae, concrete models, or gestures can support their argumentation when defending their ideas or questioning another person (the teacher or a colleague) about an aspect that is unclear for them. So, students can use non-verbal representations as an argumentative tool. This is a common practice in MBT, that is, in contexts in which students are asked to express their ideas through the production of models that can be expressed in many models of representation, as well as to test, modify, and use such models in distinct contexts.

From the recognition that teachers and students frequently use non-verbal representations to express and defend their ideas, and considering the focus of this book, the previously identified questions can be reframed and broadened: What are the roles of non-verbal representations in students' and teachers' argumentation in MBT contexts? The discussion of this question is the focus of this sub-section.

When analysing the argumentative situations that occurred in the MBT of intermolecular interactions (Mendonça & Justi, 2013b), it was realised that students and the teacher often used non-verbal representations, even when the activities did not require them to. Then, the authors produced a broader transcription of such argumentative situations by capturing all modes of representation used, that is, they included not only speech (verbal mode), but also gestures (gestural mode), concrete models (concrete mode), drawings (visual mode), and formulae (symbolic mode). Next, they analysed such a transcription by having in mind the context of each argumentative situation in terms of the modelling process. This allowed the identification of the functions of the representations in the argumentative discourse and/or in each modelling stage. As discussed in Oliveira, Justi, and Mendonça (2015), non-verbal representations were used in all context related or not to their aims in given stages of modelling.

In contexts that were independent of the teacher or students' aims at any stage of modelling, non-verbal representations were used for *reinforcing what the individual was expressing orally*, and for *substituting specific scientific vocabulary*. In the first case, both the teacher and the students seemed to have used non-verbal representations when attempting to express their ideas more clearly, or even to emphasise a specific aspect of their ideas. For instance, when a student explained that the particles that constitute a given material separate from each other when it melts, he gesticulates, miming a separation. On the other hand, the reasons for using non-verbal representations for substituting specific scientific vocabulary varied according to the individuals doing so. In the case of the students, this occurred when they did not know, or had forgotten, the appropriate scientific term referring to a process, phenomenon, or entity they have already studied, or when they did not yet feel sure about using a certain vocabulary term. This occurred several times when students: (i) answered the teacher questions by building a concrete model and only showing it to the teacher (as if the model itself could answer the question or were the answer to the question), rather than by verbally expressing such ideas; and (ii) used a previously constructed concrete model to explain the behaviour of the represented substance and said: *"This one will attract that other."* without using the word 'atom'. In the case of the teacher, this situation occurred when she considered it simpler not to use specific vocabulary in order to facilitate student understanding. For instance, during a discussion, a group of students used the word 'pairs' referring to a concrete model formed by two balls (representing the iodine molecule). Data show us that they did not use the word 'molecule' because they were not sure about both the meaning of 'molecule' and the difference between 'molecule' and 'compound'. Then the teacher continued the discussion using the word 'pairs' and mainly pointing to their concrete model. Later on, she asked other questions that helped them to understand the difference. Notwithstanding the importance of each of these purposes for using non-verbal representations, they have no clear relationship with the performance of specific stages of modelling: they were observed throughout the whole process.

On the other hand, Oliveira et al. (2015) also concluded that non-verbal representations were used for specific functions depending on the context of individual's aims. Five main function were identified:

- *To present the model showing the information that supports a proposal.*

  Students used non-verbal representations for this purpose when they expressed their ideas through a model produced in response to a question arising from an activity or a (direct or indirect) request from the teacher to reformulate a previous model, that is, when they have just created or modified their model. For instance, in one of the activities students were asked to produce models that could explain the behaviour of the iodine before, during and after being heated. When asked to communicate their ideas to the whole class, one of the students showed the models represented in Fig. 6.7 and said: "In the solid state, the atoms are bound to each other; during the heating of the system, they become separated from each ether; and after the heating they stayed together again."
- *To check understanding of a representation or a response, presenting the representation once again or stressing an aspect implicit in the group's presentation of the model.*

  This purpose was identified when the individual (the teacher or a student) (i) used a representation with the aim of confirming his or her understanding of

Fig. 6.7 Concrete models produced to explain the iodine before, during, and after being heated

something another individual had expressed. For example, when the teacher repeated an explanation that a student had just provided and used the students' concrete models to represent the movement of the molecules they'd just described in their answer in order to check her understanding of the groups' ideas; or (ii) selected a specific aspect of the presentation of an idea which was not well understood or which was implied, and made use of the representation to try to understand this aspect or make it clearer. Therefore, it was observed mainly in the stages of expressing and testing a model. For instance, one of the groups built a concrete model for graphite as a single ball, but said: *"its particles tend to separate from each other"*. Then the teacher pointed to their concrete model and said: *"What are you going to separate if there is only one here?"* This made the students rethink their representation and changed it by including more balls.

- *To explain a concept or a specific aspect of a model.*

  Both the teacher and the students used non-verbal representations with such a purpose when they give an explanation related to a given model (when responding – or not – questions made by other individuals during discussions of the ideas involved in a model). It is worth emphasising that the teacher used non-verbal modes of representation with this function fewer times than the students. However, when doing so, she established clear relationships between non-verbal models for a given entity expressed by using different modes of representation. For instance, she used formula, condensed and ball-and-stick models for iodine, and emphasised to students how they had been able to show the movement of the molecules by using their concrete models in simulations. By so doing, she contributed to enhance the students' visualisation skills.

- *To deal with some incoherence in the representation of a model and to seek a better way of representing it.*

  This purpose was identified when the representation was used to prove some incoherent aspect within it (often related to the use of representation codes which conflict with some concept involved in the representation), or in another representation to which it is related. The main example was observed during the first activity, when a group of students produced a concrete model for iodine by binding a red and a blue play dough balls. Then, the teacher asked them what they had represented, why they had used different coloured play dough, what would be different if they had used same coloured balls. She then actually produced models made of two red balls and two blue balls. When students saw the three models, they realised the incoherence in their representation.

- *To make reference to a representation that was of particular interest.*
  Non-verbal representations were used for this purpose when the individual simply pointed to a representation he or she had mentioned in a dialogue. This was observed in the stages of expressing the model, testing, or evaluating it.

Considering the purposes identified for using non-verbal representations when performing distinct stage of modelling, it becomes clear that the involvement in modelling activities foster and supported the occurrence of argumentative situations, as previously emphasised in few other publications (Latour & Woolgar, 1986; Lehrer & Schauble, 2005; Windschitl et al., 2008). When comparing the frequency with which each individual used non-verbal representations when arguing, Oliveira et al. (2015) observed that students' main purposes were *to substitute specific scientific vocabulary*, *to present the model showing the information that supports it*, and *to explain a concept or a specific aspect of a model*. On the other hand, the teacher's main purposes for using non-verbal representations were *to check understanding of a representation* and *to make reference to a representation of interest*. All of them are consistent with the context studied, that is, with the participation of students in modelling-based and argumentation teaching activities to which they were not accustomed. However, the analysis of how and what students learnt from such teaching activities shows that their involvement in argumentative situations was richer (in terms of clearly expressing their justifications) and more effective (in terms of being able to convince others of their ideas) in cases in which non-verbal representations were part of the arguments. This means that students use of distinct modes of representation when performing modelling activities can really contribute to achieving the main goals of argumentative situations (sense making, articulating, and persuading), as well as to learning science – an aspect also emphasised in empirical and theoretical publications (Hand & Choi, 2010; Jiménez-Aleixandre, 2010).

Concluding Remarks

By taking into account theoretical and empirical constructs about modelling and argumentation, as well as a set of data gathered in the context of MBT of abstract chemical topics, we were able to establish relationships between argumentation and modelling based on (i) the argumentative and modelling skills displayed by students when performing distinct stages of modelling, and (ii) the use of several non-verbal modes of representation when expressing models that supported students and the teacher argumentation or even constituted their arguments. The discussion of both sets of relationships shows that the role of argumentation in modelling goes beyond the support of discussions during the tests and reformulation of models (as emphasised in most of the current literature), or the occurrence of argumentation in all stages of modelling (as discussed in Mendonça & Justi, 2013b; Passmore & Svoboda, 2012). The relationships discussed here show that both epistemic practices influence and are influenced by each other. On the one hand, the display of specific argumentative skills can contribute to the performance of specific stages of modelling. On the other hand, (i) modelling (including the production and use of non-verbal models) supports the occurrence of interpersonal or intrapersonal argumentative context; and (ii) all the stages of modelling involves one or more of the goals for constructing scientific explanations and participating in argumentative situations that were proposed by Berland and Reiser (2009): sense making, articulation, and persuasion. This reinforce our idea that MBT is a good approach for fostering authentic science education since it may support students learning of several issues concerning two of the most important scientific epistemic practices. This encourages us to continue investigating in order to enhance our knowledge in this area, thus fostering better-founded actions in science teaching.

References

Allchin, D. (2011). Evaluating knowledge of the nature of (whole) science. *Science Education, 95*(3), 518–542.

Baker, M. (2009). Argumentative interactions and the social construction of knowledge. In N. M. Mirza & A.-N. Perret-Clermont (Eds.), *Argumentation and education: Theoretical foundations and practices* (pp. 127–144). Dordrecht, The Netherlands: Springer.

Bell, P., & Linn, M. (2002). Scientific arguments as learning artifacts: Designing for learning from the web with KIE. *International Journal of Science Education, 22*(8), 797–817.

Berland, L. K., & Reiser, B. J. (2009). Making sense of argumentation and explanation. *Science Education, 93*(1), 26–55.

Böttcher, F., & Meisert, A. (2011). Argumentation in science education: A model-based framework. *Science & Education, 20*(2), 103–140.

Bravo, B., Puig, B., & Jiménez-Aleixandre, M. P. (2009). Competencias en el uso de pruebas en argumentación [Competence in the use of scientific evidence in argumentation]. *Educacion Quimica, 20*(2), 126–131.

Cambrosio, A., Jacobi, D., & Keating, P. (2006). Arguing with Images: Pauling's theory of antibody formation. In L. Pauwels (Ed.), *Visual cultures of science: Rethinking representational practices in knowledge building and science communication* (pp. 153–194). Hanover, New Hampshire: Dartmoutn College Press.

Chin, C., & Osborne, J. (2010). Students' questions and discursive interactions: Their impact on argumentation during collaborative group discussions in science. *Journal of Research in Science Teaching, 47*(7), 883–908.

Cross, D., Taasoobshirazi, G., Hendricks, S., & Hickey, D. (2008). Argumentation: A strategy for improving achievement and revealing scientific identities. *International Journal of Science Education, 30*(6), 837–861.

Deng, F., Chen, D.-T., Tsai, C.-C., & Chai, C. S. (2011). Students' views of the nature of science: A critical review of research. *Science Education, 95*(6), 961–999.

Driver, R., Newton, P., & Osborne, J. (2000). Establishing the norms of scientific argumentation in classrooms. *Science Education, 84*(3), 287–312.
Duschl, R. (2008a). Quality argumentation and epistemic criteria. In S. Erduran & M. P. Jiménez-Aleixandre (Eds.), *Argumentation in science education: Perspectives from classroom-based research* (pp. 159–175). Dordrecht, The Netherlands: Springer.
Duschl, R. (2008b). Science education in three-part harmony: Balancing conceptual, epistemic, and social learning goals. *Review of Research in Education, 32*(1), 268–291.
Duschl, R., & Osborne, J. (2002). Supporting and promoting argumentation discourse in science education. *Studies in Science Education, 38*, 39–72.
Erduran, S., Simon, S., & Osborne, J. (2004). TAPping into argumentation: Developments in the application of Toulmin's argument pattern for studying discourse. *Science Education, 88*(6), 915–933.
Hand, B. M., & Choi, A. (2010). Examining the impact of student use of multiple modal representations in constructing arguments in organic chemistry laboratory classes. *Research in Science Education, 40*(1), 29–44.
Hogan, K. (2000). Exploring a process view of students' knowledge about the nature of science. *Science Education, 84*(1), 51–70.
Irwin, A. R. (2000). Historical case studies: Teaching the nature of science in context. *Science Education, 84*(1), 5–26.
Jiménez-Aleixandre, M. P. (2010). *10 Ideas Clave: Competencias en argumentación y uso de pruebas [10 Key Ideas: Competences in argumentations and use of evidence]*. Barcelona, Spain: Graó.
Jiménez-Aleixandre, M. P., & Erduran, S. (2008). Argumentation in science education: An overview. In S. Erduran & M. P. Jiménez-Aleixandre (Eds.), *Argumentation in science education - perspectives from classroom-based research* (pp. 3–27). Dordrecht, The Netherlands: Springer.
Jiménez-Aleixandre, M. P., & Pereiro Muñoz, C. (2002). Knowledge producers or knowledge consumers? Argumentation and decision making about environmental mangement. *International Journal of Science Education, 24*(11), 1171–1190.
Justi, R. (2015). Relações entre Argumentação e Modelagem no Contexto da Ciência e do Ensino de Ciências [Relations between argumentation and modelling in science and science education contexts]. *Ensaio: Pesquisa em Educação em Ciências [Essays on Science Education Research], 17*(Especial), 31–48.
Kelly, G., Druker, S., & Chen, C. (1998). Students' reasoning about electricity: Combining performance assessments with argumentation analysis. *International Journal of Science Education, 20*(7), 849–871.
Kuhn, D. (1991). *The skills of argument*. New York, NY: Cambridge University.
Kuhn, D. (2010). Teaching and learning science as argument. *Science Education, 94*(5), 810–824.
Latour, B., & Woolgar, S. (1986). *Laboratory life: The construction of scientific facts*. Princeton, NJ: Princeton University Press.
Lehrer, R., & Schauble, L. (2005). Developing modeling and argument in elementary grades. In T. A. Romberg, T. P. Carpenter, & F. Dremock (Eds.), *Understanding mathematics and science matters* (pp. 29–53). Mahwah, N.J.: Erlbaum.
Lemke, J. L. (1998). Multiplying meaning: Visual and verbal semiotics in scientific text. In J. R. Martin & R. Veel (Eds.), *Reading science: Critical and functional perspectives on discourses of science* (pp. 87–113). New York, NY: Routledge.
McDonald, C. V., & McRobbie, C. J. (2012). Utilising argumentation to teach nature of science. In B. J. Fraser, K. G. Tobin, & C. J. McRobbie (Eds.), *Second international handbook of science education* (Vol. 2, pp. 969–986). Dordrecht, The Netherlands: Springer.
Mendonça, P. C. C., & Justi, R. (2013a). Ensino-Aprendizagem de Ciências e Argumentação: Discussões e questões atuais [Science education and argumentation: Current discussions and questions]. *Revista Brasileira de Pesquisa em Educação em Ciências [Brazilian Journal of Research in Science Education], 13*(1), 187–216.

Mendonça, P. C. C., & Justi, R. (2013b). The relationships between modelling and argumentation from the perspective of the Model of Modelling diagram. *International Journal of Science Education, 35*(14), 2007–2034.
Mendonça, P. C. C., & Justi, R. (2014). An instrument for analyzing arguments produced in modeling-based chemistry lessons. *Journal of Research in Science Teaching, 51*(2), 192–218.
Mendonça, P. C. C., & Justi, R. (2009). *Proposição de um Instrumento para Avaliação de Habilidades Argumentativas - Parte I - Fundamentos teóricos [Production of an instrument to assess argumentative skills - Part 1 - Theoretical foundations].* Paper presented at the VII Encontro Nacional de Pesquisa em Educação em Ciências [VII Brazilian Conference on Research in Science Education], Florianópolis.
Oliveira, D. K. B. S., Justi, R., & Mendonça, P. C. C. (2015). The use of representations and argumentative and explanatory situations. *International Journal of Science Education, 37*(9), 1402–1435.
Passmore, C. M., & Svoboda, J. (2012). Exploring opportunities for argumentation in modelling classrooms. *International Journal of Science Education, 34*(10), 1535–1554.
Sampson, V., & Clarke, D. (2008). Assessment of the ways students generate arguments in science education: Current perspectives and recommendations for future directions. *Science Education, 92*(3), 447–472.
Toulmin, S. (1958). *The uses of argument.* New York, NY: Cambridge University Press.
Windschitl, M., Thompson, J., & Braaten, M. (2008). Beyond the scientific method: Model-based inquiry as a new paradigm of preference for school science investigations. *Science Education, 92*(5), 941–967.
Zohar, A., & Nemet, F. (2002). Fostering students' knowledge and argumentation skills through dilemmas in human genetics. *Journal of Research in Science Teaching, 39*(1), 35–62.

Chapter 7
The Contribution of Visualisation to Modelling-Based Teaching

Abstract Both the creation of models and their communication to other people involve visualisations. These are, respectively, 'internal' (or mental) and 'external' (or public) representations, with the latter confusingly also being called visualisations. Perceptions by one of the five senses provide external representations. The modes of external representation of particular importance in science education are the: gestural, concrete, static visual (pictures, diagrams, graphs, mathematical and chemical equations), dynamic visual (drama, animation, simulation), oral and auditory. The skills and abilities that constitute meta-visual competence in the modes are reviewed in this chapter, for they enable the central element of modelling – the design and conduct of thought experiments – to take place. Consequently, the skills and abilities of both modelling and of visualisation are mutually developed and employed during MBT.

The Growing Importance of Visualisation

Since the dawn of 'the age of science', arguably about 1400 CE, evermore detailed phenomena in the world-as-experienced have been studied, producing increasingly complex and detailed models of their behaviour and constitution. In order to be accepted as contributing to scientific progress, these models then have to be communicated by their creators to other scientists for critical review and, given sufficient support, for general publication in scientific journals and use in other research. Creating, seeking approval for, and disseminating these models inevitably taxes the imaginations of scientists, tasks for which every help is no doubt much appreciated. With the somewhat parallel growth of systematic science education, in Europe and North America from about 1850, one problem that rapidly emerged was how to successfully promote an understanding of the ever-expanding range of established models in students. More recently, in order both to support the development of more scientists and to facilitate greater creativity, as well as providing insight for the general public into how models are devised and evaluated, a greater emphasis has been placed in science education on the skills of modelling. Visualisation plays a central and increasing role in all three of these activities: the creation of models, their evaluation by the scientific community, and their communication to students of science of all ages.

J.K. Gilbert, R. Justi, *Modelling-based Teaching in Science Education*, Models and Modeling in Science Education 9, DOI 10.1007/978-3-319-29039-3_7

A visualisation, also commonly called 'what is seen in the mind's eye', is a colloquial way of referring to the result of the formation of a 'mental image' or 'mental model'. Visualising, the formation of a visualisation, is a quasi-perceptual experience, in that it resembles perceptual experience, but is one which can occur in the absence of the external stimuli (Thomas, 2014). The steadily increasing social pressure to create and communicate visualisations has gone hand-in-hand with an expansion of the availability of support for doing so. Whilst early efforts were largely confined to sketches and line drawings, the advent of high-quality reprographics and lately the introduction of colour into printing, has expanded the capability of the manuscript, book, overhead projector transparency, or handout, to include 'illustrations' of all kinds. The ready availability of readily mouldable plastics since the 1930s has enabled a wide range of physical models to be produced quickly and cheaply. However, it is the exponential growth in the number and capabilities of the personal computer, and lately of mobile phones and tablets, that have had by far the greatest impact on all creation, dissemination and use of models. This expansion of the ways of producing, depicting, and communicating models has inevitably been associated with a more careful consideration of what those acts involve. Whilst that consideration is still far from complete, progress has been made: the notions of 'visualisation' and 'representation' have emerged and must be distinguished between.

The Notions of Creating and Representing Visualisations

All models are created in the minds of individuals: these are usually known as 'mental models' and retained as 'visualisations'. Attempting to convey, to another person, the information included in a mental model involves producing a representation of the visualisation of it that can be interpreted by use of one or more of the five senses. Such a representation, when made available to others in the public domain, and therefore able to be interpreted by each individual in that domain, may be called an *external representation*. While the creator of an external representation will intend that the meanings that are embodied in it are understood by all those who perceive it, this ambition is all too often not realised. For example, for every word used in science, a range of interpretations are acquired and used by students: the so-called 'alternative conceptions' (Gilbert & Watts, 1983). The meanings attached in the brain to a given external representation may therefore be called a personal *internal representation* of it and is a visualisation of it.

Because sight plays such an important role in cognition generally, the word *visualisation* is often used, confusingly, instead of the more general term *external representation*. However, as the mechanisms of meaning-making in respect of internal and external representations are thought to be very similar if not identical (Kosslyn, Pinker, Smith, & Shwartz, 1982), the word visualisation is often *also* used in the literature to mean internal representation. As the mechanisms of meaning-making for internal and external representations are so closely related, we will use the word *visualisation* to mean either of them, with the context of a statement providing guidance on which is intended.

Visualisation, whether internal or external, operates at three levels. The first is at the *macroscopic* level, that of the phenomenon being studied, as it is directly perceived. Examples are: a solution of a chemical; a puck moving on a low-friction surface; the cross section of a leaf. The second is at the *sub-microscopic* level, a representation of those entities that are thought to underlie the properties being displayed. To continue with the three examples, the sub-microscopic levels involved are respectively: molecules or ions in solution; lubricants between the object and the surface on which it moves; plant cells. The third is the *symbolic* level, where quantitative abstractions are used to depict the entities at the sub-microscopic level. Thus, respectively: chemical equations and the mathematical equations are used to calculate the amount of substance; mathematical equations used to depict movement; mathematical equations are used to depict the type, number and position of cells (Gilbert, 2008; Johnstone, 1993).

As representing the external world is far easier than doing so for the internal world, it is usual to talk about internal visualisation in terms of the modes of representation used in external representation, although each of these, of course, has an internal consequent. In discussing the modes of external representation, we will identify their especial value in visualising the levels of representation.

Modes of External Representation

The five senses – sight, hearing, smell, taste, and touch – provide all our perceptions of the world-as-experienced. In terms of the external representation of scientific models, taste and smell make very limited contributions. This is so because the senses of taste and smell have eluded valid and reliable analysis and so are little used in science. The sense of taste is mostly associated with the phenomena of food and drink, for example with the appreciation of a particular vintage of a wine, where smell and taste seem to be cognitively related. The sense of smell is seldom encountered, except for example in the evocation of a particular emotion that a cosmetic perfume is intended to produce. Those remaining – touch, sight, hearing – provide the three generic modes of external representation, whilst their perceptual significance is so great that each has produced a range of sub-modes. Each of these modes and sub-modes enables particular aspects of a model to be emphasised, and by implication de-emphasised or omitted, hence having a distinctive code of representation relative to the parent internal model. Moreover, these different modes and sub-modes often produce external representations by working simultaneously or sequentially on existing internal representations. This richness means that tidy, convincing, taxonomies of the nature and capabilities of modes and sub-modes are difficult to produce.

In the next three sections, we outline some of the main ideas about each mode (and their associated sub-modes), assuming that they are the ones used by students and teachers to produce external representations and, most probably, internal representations. The issue addressed here is therefore 'what particular properties of the various modes and/or sub-modes cause them to be selected as ways to produce an internal representation and hence an external representation?'

Modes of External Representation Based on Touch

This group consists of two types of mode: one in which the touch is implicit but not actual, that is it is based on the production of gestures; the other in which touch is explicit and actual, it is based on the use of concrete objects.

The Gestural Mode

In explaining something to another person, for example when giving directions, almost all of us make use of gestures with our hands and arms. We usually do so without being aware of the act. This lack of awareness can be physically risky, for the meaning and social acceptability of particular gestures vary a great deal between cultures! Not surprisingly then, the use of the gestural mode of representation in the communication of science and science education is under-appreciated, under-valued, and hence under-researched. To some extent, this under-valuation probably arises from the fact that such gestures are normally accompanied by speech and their significance is perceived to be subordinate to that speech. Nevertheless, gestures do seem capable of amending or reinforcing the meanings intended by words that are uttered at the same time. The value of gestures may therefore be the capacity to identify and emphasise the emotional commitment, either positive or negative, to what is being depicted by the speech of a person making a representation. For those having an impairment of hearing, gestures are valuable: for the deaf, 'sign language' is truly invaluable.

Whilst they can be made with any part of the body (eg shrugging the shoulders, averting the gaze), the variety of gestures most commonly used in the teaching of science is produced by the movement of hands and arms. McNeill (2005) produced a classification of gestures along several non-exclusive dimensions of which the four most readily evident are:

- *Deixic gestures*. These involve pointing at a real or imagined object. For example, moving ones hands to invoke the spiral shape of the DNA molecule;
- *Metaphoric gestures*. These involve an abstraction being referred to by means of gestures that liken it to something that is familiar. For example, a force is referred to by a gesture indicating its line of action;

- *Iconic gestures*. These emphasise the semantics of speech, the meaning that is being conveyed: that some change takes place in an object or event, for example a change in the climate.
- *Temporal highlighting gestures*. These gestures are used to emphasise the significance of time or the number of occurrences of what has been said. For example, to suggest the evolving impact of climate change on crops over some years.

In the absence of a body of research into the use of gestures in science education, it is not currently possible to say, with any surety, which types of gestures will be especially valuable in support of representations at each of the three levels (macro, sub-micro, symbolic).

The thing that such gestures have in common, perhaps inevitably, is that they are used in respect of topics for which it is difficult to form internal representations. Thus Padalkar and Ramadas (2011) focused on the role of gestures in the learning of elementary astronomy (the Earth-Moon-Sun system) in Grade 8 classes, Herrera and Riggs (2013) examined the use of gestures when US undergraduates learnt about stratigraphy in geology classes, and Roth and Welzel (2001) looked at the relationship between gestures, laboratory practical work, and conceptual learning for Grade 10 students of physics. These studies draw on a similar theoretical background and reach a consensus about the value of gesture as a mode of representation. The notion of 'embodied cognition' (Wilson, 2002), the idea that bodily movement is reflected in cognition and vice versa, has been found to be helpful in understanding the cognitions that are associated with the perceptual consequences of gesture. In other words, the embodied cognition that results from gesture is metaphorically reflected in the resulting schema that provide internal representations (Lakoff & Johnson, 1980). The use of deistic, iconic and metaphoric gestures were found in all the studies of learning cited above, albeit to different degrees, whilst the existence of a link between the use of gestures and associated speech has been confirmed by Goldin-Meadow (2006), and the particular value of gestures in reasoning about mechanical systems has been established (Kastens, Agrawal, & Liben, 2008). It does seem that gestures in fact play a central role in all concept formation and change (Givry & Roth, 2006), perhaps because of the affective drives associated with those events. The growing recognition of the importance of the affective aspects of learning will perhaps be an incentive for giving gestures a higher saliency in enquiries into the operation of the modes of representation.

The Concrete Mode

Some concrete (aka 'material') representations are of the type that Rom Harré called 'homeomorphs'. These are produced in a material that is either the same as or similar in appearance to that being represented and are typically larger or smaller than the original (Rothbart, 2004). A demonstration 'model' of a proposed oil refinery and a toy motorcar are respectively examples of this type. Many concrete representations are of the type that Rom Harré called 'paramorphs' (Rothbart, 2004), in which the representation and that which is represented are of different material kinds, and are related analogically to each other. A wide variety of sub-modes of paramorphic concrete representation exist, facilitated by the many materials that can be used in their construction. A typical example in chemistry is the ball-and-stick representation of the shape of an individual molecule produced in polystyrene. One in biology is the 'cut away' representation of the veins/arteries relationship in the blood circulation in a body, produced by the use of mouldable plastics. In physics, a 'blown-up' representation of a micro-electronic circuit made of wire enables the functions of the parts of the original to be readily discussed. A wide variety of sub-modes of paramorphic concrete representations exist and are produced commercially. These are usually protected by patent laws and are marketed for science class use in many countries. No comprehensive database of them, even on a national base, seems to exist, relevant work being somewhat dated (Savory, 1976) or merely illustrative (Francoeur, 1997) and mainly concerned with chemistry.

A recent development in the field is that of '3D printing' (3D Printing, 2015). In this technique, which evidently has many applications, a 3D scan is made of an object using a 3D modelling programme and entered into a 'Computer Aided Design File' from which a 3D representation of an object can be printed using a range of materials.

Modes of External Representation Based on Sight

Given the overwhelming importance of sight in human perception, it is not surprising that a large range of visual representations have been developed. These have been subjected to research from a wide range of theoretical perspectives. The outcome of this is the absence of a generally agreed taxonomy of them (Blackwell & Engelhardt, 1998). In these circumstances we can fall back on a pragmatic taxonomy based on work in the science education community (Eilam, 2012; Gilbert, 2011). In general terms, sight-based representations may be divided into static visualizations and dynamic visualisations. The static group includes photographs, diagrams, charts and maps, graphs and data tables. The dynamic group includes films/ video, animations/simulations. In the necessarily concise treatment of these diverse modes of external representation, we have built on the summary of the field by Eilam (2012).

Static Visual Representations

1. Pictures

The digital camera, the mobile phone, and the tablet have together made the production and distribution of photographs a widespread practice in recent years. They have several distinct contributions to make to science education. Firstly, they can focus the learner's attention on specific elements of any accompanying text. Secondly, pictures can be designed so as to summarise distinct and lengthy segments of that text. Thirdly, they can provide a record of any object without any distortions of its shape other than those inherent in the creating of boundaries (the edges) of the picture. The organisation of the elements contained within those boundaries is inevitably emphasised so that the relationship between those elements is therefore clarified. Taken together, these characteristics collectively support the value of photographs in the formation of mental models (Levin & Mayer, 1993), especially those of the macro level of phenomena that are not readily available, for example those of the temperate zone landscape at the heights of the four seasons.

Photographs usually make these contributions in relation to associated text. They can have any one of a range of roles as indicated by the nature of the accompanying captions. They can be: decorative, having no worthwhile relation to it; representative, in effect repeating part of the text but without adding anything to it; organisational, in that they can provide a structure of relationships for the content of the text; interpretational, in that they clarify any difficult elements in the text; transformational, in that they can aid memorisation of the text by providing a visual mnemonic for it (Levin, Shriberg, & Berry, 1983). The careful selection and precise placement of photographs within a text must reflect clarity of contribution and role if they are not to confuse the student (Willows, 1978). However, it does seem that the potential of photographs is often not fully exploited because of a failure to exploit the explanatory possibilities of the various roles (Pozzer & Roth, 2003).

2. Diagrams, charts, maps

The meanings of the words 'diagrams', 'charts', and 'maps' are closely related, a situation made even more complicated by the existence of a wide variety of sub-modes within each type, for example 'flow diagrams', 'Venn diagrams', 'concept maps' all fall within the general framework of 'diagram'. These sub-types of 'diagram' provide qualitative information by the use of universal symbols, for example lines and arrows, to link pictures of varying complexity that represent parts of an entity. Concept maps represent organisational structures and relationships between classes of entity, for example when depicting hierarchies. Maps provide quantitative information regarding the arrangement of real objects in actual space, for example, distances, magnitude, distribution (Winn, 1991).

3. Graphs

The variety of sub-types of 'graph' are equally legion, all of which are used to present quantitative information. Of these, the categorical line graph and the continuous line graph play major roles in science education. Kosslyn (2006) identified the three components that are needed to produce such graphs. The first is a 'framework' that states the kind of data presented, is it to be 'categorical' or 'continuous, and of its distribution as an independent variable ('x axis') and dependent variable ('y axis'). The second is the 'content', the distribution of corresponding values of the two variables. The third is the 'label' of the graph, the scale of units provided by the axes and the overall title of the graph. Of the various sub-types available, bar graphs more effectively convey information about rank orders, line graphs show trends to best effect, whilst pie charts (so called) show the relative proportions of the components of a whole.

4. Data table

Data tables are usually the intellectual step prior to the construction of a graph. The changes in the two variables are listed chronologically (or in any other order) within the tables.

5. Mathematical models

There seems to be no agreed definitions of mathematical models and modelling (Ferrucci & Carter, 2003). A convenient way through the variations available may simply be to view a mathematical model as the representation of some aspect of the world-as-experienced by the use of mathematical techniques in such a way that predictions can be made about it. The techniques used in producing such representations have evolved over many years, as the case study of the evolving mathematical description of the 'trajectory of a cannonball' shows (Swetz, 1989). The aims, form, and degree of address to mathematical models and their production in the general curriculum of school mathematics education has also changed over the years (Niss, 1987). Until recent years the main emphasis on mathematical modelling has been on its use in teaching the applications of mathematics in general (Burges, 1980) and in the teaching of fundamental concepts in science in particular, for example in physics (Paton, 1990), in chemistry (Paton, 1991a) and in biology (Paton, 1991b). However, there is some evidence that mathematical modelling, as a mental process, supports conceptual development in general.

The acceptance of the increased role of models and modelling in the teaching of scientific methodology (Justi & Gilbert, 2002) have given teaching of the technique greater saliency in school-level science education recent years. This has been given added impetus because as mathematical modelling plays a major role in the creation and communication of scientific knowledge, all university students of science and technology must become conversant with its techniques, of which those involving statistics are especially prominent.

6. Chemical equations

The other major class of representation at the visual type is the 'chemical equation'. The version of chemical equation that must be used in order to symbolically model a chemical system for validation by professional chemists has, by international convention (IUPAC, 2014), to conform to a number of rules – it must adhere to a specific 'code of representation'. Thus: only chemical species that change during a chemical reaction must be included in it; those species must all be depicted using the standard system of symbols for elements and conform to the 'law of the conservation of mass'; the chemical status (ions, molecules, free radicals, etc.) and the physical state (solid, liquid, solution, gas) of those species must be shown; the possibility of the reaction being reversible, to any extent, must be shown; enthalpy changes may be shown. To take a simple case: the reaction of potassium hydroxide and hydrochloric acid in aqueous solution, a reasonable representation would be:

$$OH^{-}_{(aq)} + H^{+}_{(aq)} \rightleftarrows H_2O_{(l)}$$

This issue of representational convention is important, for other historically validated conventions are still in use, for example in old textbooks. For example, there is the 'word equation'. In the above case this would be:

$$\text{Potassium hydroxide} + \text{hydrochloric acid} \rightarrow \text{potassium chloride} + \text{water}$$

As Taber (2002) points out, this extreme type of case violates all the elements of the international convention. Of course, some commonly used representations only violate some of the conventions, for example:

$$OH^- + H^+ \rightarrow H_2O$$

However, the teaching of the standard international convention to all students is necessary if they are to later undertake more advanced studies of chemistry.

Dynamic Visual Representations

As the sub-section heading implies, dynamic visual representations are those that place considerable reliance on sight and depict any movement taking place. Four forms can be identifies: drama, video, animations, and simulations.

1. Drama

Although drama makes extensive use of gestures, it is very different from that genre in two ways: firstly, what takes place is conscious and deliberate on the part of the 'actors'; secondly, the activity is sustained, not transient, and repeated. Dorion (2009), who has reviewed the field and written about his own experiences in science classrooms, defines it as "role play within an imagined situation and enacted within the human dimension" (p. 2249).

Although it is not yet apparently widely used, drama has been used in science classrooms. For example: to role-play historical episodes in science, to depict the forensic investigation of imaginary crimes, and to model events of scientific interest like the behaviour of electrical circuits and the function of the kidneys (Dorion, 2009). Staged within a Vygotskian dialogic environment, role-play in these very different contexts has been found to promote student learning in the topics under consideration by emphasising their relevance to the lives of the students.

2. Video

The recent and rapid rise in the availability of low cost, small, robust, video cameras must surely lead to their increased use in science education generally and, in particular, for keeping a record of modelling projects. As this exploration evolves, some current use is being made of the simultaneous use of the amalgam of visual and auditory signs present in video as students are presented with surrogate access both to inaccessible phenomena and to the processes of modelling these, for example as shown in the BBC *Horizon* series of programmes.[1] The creative use of video cameras does impose a cognitive load on students, for example through the use of colour, the complexity of the information to be presented, and the inevitable speed of that presentation. Despite these demands, students must be able to focus their attention on salient features of the presentation when either viewing a pre-recorded video or when preparing their own. These challenges – which are often considerable – can be addressed by the careful preparation of the class for the use of the video and by the repetition of its use until it has been fully understood (Eilam, 2012).

3. Animation

Animation involves the actual portrayal of any changes that take place in a phenomenon over a period of time where these take place either too slowly or too rapidly to be captured on video. This simplification of the time element enables the core aspects of a phenomenon (the entities that it consists of, the relationships between them, the changes that take place in the phenomenon) to be focused on, for distraction is minimised. As an animation is a representation of reality, the approach does not usually enable the user to interact with the animation, other than to change the speed of representation of the reality depicted.

4. Simulation

The notion of simulation is closely allied to that of animation, the difference being that the object represented is a model rather than the reality itself. This crucial difference does enable the user to extensively interact with the simulation by altering the nature and degree of representation that is the basis of the model. A rapidly expanding field in science education, the current state of the art has been recently reviewed (Landriscina, 2013).

The widespread availability of personal computers with very large memory capacities is leading to the increased use simulations in science education at all levels, not least in chemical education (Dori, Rodrigues, & Schanze, 2013). It has thus been found possible to produce computer-managed pseudo-3D versions of concrete models, often called 'virtual representations', in which not only is it possible to manipulate the composition and shape of molecules but in which simulations of chemical and biochemical reactions can also be run. The attractions of such systems are that widespread and sustained use of them can be readily provided. Whilst access to the actual concrete representation has much to commend it (see earlier in this chapter), students do readily learn through the use of simulations and virtual representations (Savec, Vrtacnik, & Gilbert, 2005; Urhahne, Sabine, & Schanze, 2009).

---

[1] Horizon is a BBC series exploring topical scientific and philosophical issues and their effects for the future, which has been broadcasted since 1964.

Modes of External Representation Based on Speech and Hearing

Hodson (2009) has commented, in respect of the language of science, that

> Students will not just 'pick up' this complex language unaided. It has to be taught, practiced, deployed in authentic situations and evaluated in action, such that students come to see themselves as members of the scientific community, or the school version of it. (Hodson, 2009, p. 243)

Modelling, as advocated in this book, is an archetypal authentic situation for such learning and practicing to take place. The notion of genre, normally used to distinguish between the audiences and purposes for writing (Hand, Prain, Lawrence, & Yore, 1999), does seem capable of extension to cover the realm of speech and hearing. When modelling is widely practiced in school science, it may then be possible to tease out the main characteristics of speech and hearing which can be regarded as the intimately related components of a particular mode of representation. The discourse between students and with the teacher during the practice of modelling will certainly be of the 'interactive/dialogic' type (Scott, Asoko, & Leach, 2007) and will form the basis of such a genre.

In the meantime, we can immediately identify 'the practice of argumentation' and 'the use of analogical reasoning' as central components of a speech/hearing genre. For that reason, we devote Chaps. 6 and 8 respectively to their use. As a good understanding of their core concepts is a prerequisite for creating representations in other modes, the problems associated with their learning and use will certainly have to be faced. Thus the altered use in science of words that also have an everyday currency, for example 'ideal', negligible, 'sequence' will come into play (Cassels & Johnstone, 1985), as will the importance attached to logical connectives (Wellington & Osborne, 2001). Whilst many other issues will gain prominence as verbal language is used in the practice of science education generally, they are not unique to discourse during modelling and hence cannot be addressed here (Hodson, 2009).

All the modes of representation summarised in this section depend on the use of visualisation, both internal and external. We must therefore discuss the complex issue of the skills that visualisation entails.

The Skills Entailed in Visualisation

What are the skills of visualisation that play such a leading role in modelling-based teaching? Here again we have to comment that this field is bedevilled by lack of clarity over terminology and hence the nature of valid and reliable research: Thomas Kuhn might have said it was still in a 'pre-paradigmatic' phase (Kuhn, 1996). One consequence has been that:

> research on the structure of spatial abilities has been largely a-theoretical and has all but ignored individual differences in important aspects of spatial cognition, such as the processing of dynamic visual displays (Hegarty & Waller, 2005, p. 159).

Despite a lack of consensus, definitions of a number of dimensions of visualisation recur in the psychology and education literature. Those that have relevance to internal representation in the act of creating a model include 'spatial visualisation' which is "The ability to manipulate, rotate, twist or invert objects without reference to one's self" (Hegarty & Waller, 2005, p. 159) whilst 'spatial orientation' involves

> the comprehension of the arrangement of elements within a visual stimulus pattern and the aptitude to remain unconfused by the changing orientation in which a spatial configuration may be presented (Hegarty & Waller, 2005, p. 127).

Apparently accepting these dimensions, a recent overview of the visualisation research literature by Newcombe and Stieff (2012) tentatively identified 'six myths about visualisation': these might distort the practice of visualisation in modelling were they to be acted upon uncritically. Their rebuttals of these myths are that:

- Whilst here is some evidence that some individuals show a particular proclivity for visual representation, the use of visualisation in learning is a universally held capability.
- There is some evidence that males are better than females at visualising whilst performing certain mental tasks. However, this tendency may be an artefact of the different patterns of experience of the sexes and of the way that visualisation is assessed. In short, it should not be used as a reason to downplay the use of visualisations in the education of females.
- There is no conclusive evidence that any differences between the sexes in respect of visualisation that do persist have been genetically caused. It seems that here, as so often in education, the practices of nurture outweigh any possible effects due to nature.
- Spatial ability is not fixed, for it can be enhanced, for both sexes, by suitable education/training (as is discussed briefly later). This supports our assertion that suitable modelling activities should enhance the visual capabilities of both sexes.
- The notion that students can only engage in visualisation when they have reached a certain personal mental 'level of development' has been shown to be untrue, for very young children have been found to be able to visualise, given a task that engages them. Modelling activities therefore have a place throughout the years of schooling if the contexts being modelled are of interest.
- Although the expression of spatially-related ideas varies between languages, there is no evidence that such linguistic differences control the ability to visualise as such. The differences between languages about the meaning of the word 'visualisation' just reflect the cultural awareness of it that is being expressed. Culturally appropriate modelling activities are suitable for education in all languages.

The conclusion that can be drawn from this sub-section is that everyone is capable of both spatial visualisation and spatial orientation. Given its central role, it is therefore vital that visualisation is recognised in the educational development of the skills of modelling, and that the capabilities of students are overtly developed. At what then should we be aiming?

The Notion of 'Meta-Visual Competence' as Attainable by All

When fully capable in respect of visualisation, a student may be said to display 'meta-visual competence', and it is that at which science educational systems should be aiming. During the process of becoming meta-visually competent, a student will:

- acquire knowledge and skill in respect of all the sub-modes of visualisation likely to be encountered during the modelling of a phenomenon of interest to science education;
- become able to justify why a particular mode of representation should be used for a given purpose in respect of a given concept or model;
- become able to integrate the understanding of a given concept or model that is provided by the use of several modes of representation;
- become able to show the relationship between the three levels of representation (macro, sub-micro, symbolic) for a given phenomenon by the use of particular sub-modes of visualisation. For example, the use of different concrete visualisations to show the structural position of the heart in the body and the internal structure of the heart, together with a diagram to show the metabolic pathway of blood flowing in the body to and from the heart;
- become able to 'translate' the visualisation provided by one mode or sub-mode into that provided by another. For example, to use concrete representations and diagrams to show the composition of a crystalline solid in respectively three- and two- dimensions;
- become able to construct a visualisation for a particular purpose within any sub-mode that is capable of representing any chosen aspect of a phenomenon. The 'heart' example, given above, would be one such case;
- become able to use visualisations that have been helpful in modelling one phenomenon in order to model another. To take an historical example, the function of the pump used to drain water from mines was used by William Harvey to provide a useful representation of the function of the heart (Harvey, 1993).

The development of meta-visual competence will be gradual, for it involves acquiring a wide range of knowledge and its manifestation in the evolution of the skills of modelling. The complexity of these processes is difficult to depict for the purposes of identifying ways of supporting their development. Several aspects of such competence probably advance in parallel. It is convenient, if only for ease of communication, to represent the evolution of the various skills as taking place in distinct 'levels'.

For example, Kozma and Russell (2005) identified five distinct and successive phases – what they termed 'levels' – in the development of meta-visual competence in respect of diagrammatic forms of representation by undergraduate chemistry students in the USA. In:

- phase 1, students visualise a physical phenomenon solely as an iconic depiction of the original. They represent what they literally can see;

- phase 2, students include some symbolic elements in such an iconic depiction. Some aspects of the original are shown to symbols;
- phase 3, students include more symbolic elements in their representation, but these are not necessarily scientifically accurate;
- phase 4, students only used symbolic elements to produce a visualisation and do so correctly. They were also able to transfer a given visualisation into another mode;
- phase 5, students were not only able to produce symbolically accurate visualisations but were able to justify their decisions.

Whilst the general idea of 'phases' is potentially of educational value (if not taken too literally), research is needed into their occurrence in other subject fields. If and when the composition of 'phases' receives widespread validation, it should be possible to use the idea to structure activities deliberately designed to lead to meta-visual capability. The attainment of these 'phases' will be dependent, in ways not understood at the moment, on the interplay between suitable modelling activities, which we have already outlined, and a student's general intellectual development (Newcombe & Learmonth, 2005). The need to understand and to support these processes is necessary given the importance of visualisation in modelling-based teaching.

The Contribution of Visualisation to Modelling-Based Teaching

As we have said, the extent to which meta-visual capability is attained and deployed by students will govern the overall effectiveness of modelling-based teaching. This becomes manifest as 'thought experimentation' is engaged in by students, for this is the main vehicle in which modelling and visualisation are brought together.

The Nature of Thought Experimentation

As a broad definition, a thought experiment is the design of an activity intended to be carried out mentally either to test the validity of an explanation proposed for a natural phenomenon and/or to develop an argument that will convince others of that validity (Brown, 1991; Gilbert & Reiner, 2000).These are situations where the question 'what if?' is asked of phenomena. According to Wilkes:

> "To answer such a question, the thought experimenter needs to imagine a possible world in which the 'what if' situation actually occurs. Then the results and implications of the 'what if' situation are examined. The results may support or weaken the tested claim." (Wilkes, 1988, p. 2)

The literature on thought experiments sees them as able to take place in one of two situations (Brown, 1991). First, where the thought experiment is actually impossible to carry out in the laboratory, for example Einstein's thought experiment that involved a hypothetical ride on a beam of light. Second, where the experiment could possibly be carried out in the laboratory or elsewhere, but this is not done for some reason, perhaps because of shortcomings in the equipment available, for example Poisson's thought experiment which attempted to disprove the wave theory of light.

To accommodate the diversity of thought experiments used in science (and hence in science education), Brown (1991) proposed a threefold classification scheme for them.

- The 'Destructive Thought Experiment' is the first type. The intention here is to either destroy, or at least weaken, the acceptability of an existing or proposed explanation for a phenomenon, for example that of the Schrodinger 'Cat' situation.
- The 'Constructive Thought Experiment' (CTE) is the second type. For Brown,

  > The burden of any constructive thought experiment consists of establishing (in the imagination) the thought-experimental phenomenon. This phenomenon then acts as fairly conclusive evidence for some theory. (Brown, 1991, p. 45)

  He distinguishes between a 'Mediative' CTE, which involves using a specific well-articulated theory, for example that used in respect of Maxwell's 'Demon' in thermodynamics; a 'Conjectural' CTE, where an imaginary phenomenon is established and then an hypothesis is produced to explanation, for example in respect of Newton's Bucket; and lastly the 'Direct' CTE which is an hybrid across the mediative and conjectural types, for example for Stevin's 'Inclined Plane' which "starts with an unproblematic thought-experimental phenomenon and ends with an (explanatory) theory" (Brown, 1991, p. 41).

- The third major type is the 'Platonic Thought Experiment' from which acts simultaneously in a 'Direct' and a 'Destructive' capacity, for example in respect of Galileo's 'Free Fall Experiment'.

When modelling-based teaching is widely established and researched into, it is likely that all these types of thought experiment will be found to be in use in science education.

There has been much debate amongst philosophers about the question 'Are thought experiments scientific?' Sorensen (1992) has drawn all these arguments together in the chapter on 'Scepticism about thought experiments' of his book 'Thought Experiments'. To 'Cut the Gordian Knot' on this matter, we will simply assert that they are included in the 'tool bag' of science (and hence of science education) on the pragmatic grounds that they have been and are one of the cognitive processes widely used and valued as ways of providing explanations found to be worthwhile.

Where experiments are carried out empirically, for example in the laboratory, discussing the conclusions that arise from them is relatively easy, for all the evidence is, literally, readily to hand. Decisions on the value of explanations are thus often fairly readily arrived at in respect of empirical experiments. It is a matter of evaluating whether the evidence is accurate, repeatable, and relevant to a decision about the proposed model of the phenomenon. However, this is not the case for thought experiments, for considerable argumentation usually takes place to establish their validity: we discuss the place of argumentation in thought experimentation within the context of modelling-based learning in Chaps. 4 and 6. A starting point for such argumentation must rest on a clear view of the role of visualisations in thought experiments.

The Place of Visualisation in Thought Experimentation

A thought experiment is an internal representation, a visualisation, consisting of a number of imagined components. Reiner (1998) considered that any thought experiment could be considered as having five components:

- the posing of a problem in the form of a question to be asked in respect of the world-as-experienced;
- the creation of an imaginary world – a model of the world-as-experienced – in which it was possible to answer the question posed;
- the mental design and conduct of an imaginary experiment in which data relevant to the proposed answer to the question is collected;
- the logical analysis of that data;
- the drawing of a conclusion as to whether the visualised world and the visualised experiment have produced an acceptable answer to the question posed.

The successful conduct of a thought experiment thus depends on one being able to visualise an imagined world and to design and conduct a visualised experiment. Learning how to design and conduct an experiment, whether imaginary or empirical, is beyond the scope of this book.

We have argued above that the ability to engage in thought experimentation and the ability to engage in visualisation are mutually interdependent. The more readily students can engage in thought experimentation, the more readily they can create visualisations. The closer they are to demonstrating meta-visual capability, the more readily they can construct and 'run' thought experiments. Being frequently and actively engaged in modelling-based activities will develop both the capabilities to produce thought experiments and to produce visualisations. This must surely argue for the inclusion of modelling-based teaching at all levels of the science curriculum, if the development of capabilities in respect of visualisation and modelling are to develop synchronously. We await the testing of these assertions and, in the meantime must assume that they are justified and take action accordingly.

The Mutual Development of the Skills of Visualisation and of Modelling

The General Development of Them Both

We can readily identify some general elements of modelling-based teaching that should lead students to acquire a greater degree of capability to visualise. These are the provision of opportunities:

- to construct visualisations as frequently as possible (Prain & Tytler, 2012). The more modelling-based teaching that students encounter, the more opportunities to learn the skills of visualisation they will experience;
- to encounter several modes or sub-modes of representation when learning about all topics, especially those that are used in all the three levels of representation (macro, sub-micro, symbolic);
- to become familiar with the use of the codes of interpretation that govern the relationships between the modes and sub-modes of representation and what is being modelled;
- to encounter visualisations of which the distinctive natures have been emphasised by the use of sharp edges, distinct shapes, the use of shading and of colour, for these measures make them easier to remember;
- to 'reflect' and 'rotate' visualisations wherever this is possible during thought experimentation (all points after Tuckey & Selvaratnam, 1993);
- and, above all, to implement the pedagogic conditions for successful modelling, for example appropriate choice of task, individual and group activity by students, the teacher to adopt the role of tutor.

However, the use in modelling of each the different modes and sub-modes does provide a particular opportunity for the development of an awareness of their contribution to visualisation. Of those that we have listed earlier, there is not enough known about some, from research specifically in science education, for clear recommendations about their use to be made: gesture, pictures, and video fall into this category. Other modes and sub-modes are part of the existing general curriculum in mathematics education, such that their detailed treatment in science education using different codes of representation might well be dysfunctional: graphs and data tables fall into this category. The contribution of the other modes and sub-modes to the development of visualisation can be taught about either adjacent to or within modelling-based education, albeit with differing degrees of confidence based on research evidence for so doing. These are: concrete mode, diagrams, mathematical modelling, chemical equations, drama, animations and simulations.

Developing Visualisation Through the Use of the Concrete Mode

The great attraction of the concrete mode of representation is that the 3D natures of entities depicted are retained. It is most commonly used in the early years of school chemistry and school biology. In the former, the emphasis is on the spatial natures of both individual molecules and of aggregates of them in crystals. In the latter, the emphasis has been on the depiction of individual organs and their spatial location in living entities, although lately it has shifted closer to chemistry with the emphasis being placed in genes and their action.

In recent years, the advent of the desktop computer and tablet has led to the ready availability of them to all students, albeit in pseudo-3D versions. The summaries of research into the use of concrete representations are both dated (that is, from the pre-computer age) and focused on chemistry (Francoeur, 1997; Tuckey & Selvaratnam, 1993). In a review of the evolution of 'molecular models' in chemistry, Francoeur (1997) divides the genre into two: the 'closed' representations (otherwise called 'space filling') which has the entities depicted a being solid such that their immediate shape is shown; in the 'open' representations the entities are depicted as points, with rods or lines being used to emphasise their spatial relationship to each other. Both sub-genres enable geometry and numerical mathematics to be used to analyse them and hence to compare their representational capabilities.

Where actual concrete representations are available, students will be able to develop their visual capabilities by a structured experience of their use. Where (as is increasingly common), only computer-based pseudo-3D versions are available, students will need to be introduced to the conventions of such forms. In particular, these are the use of various cues to show: the overlap of entities, the foreshortening/extension of lines to show below-surface/above-surface inclinations, the distortion of spatial angles between entities, the relative size of the entities being depicted. Given the widespread use of visualisations in the myriad documents commonly experienced through desk-top/lap-top/tablets, it seems very likely that many, if not all, students, will have acquired a basic 'visual vocabulary' before they even begin to formally study science. This seems especially likely where students have encountered the simulations that form the basis of the myriad 'computer games' now available.

Developing Visualisation Through the Use of Diagrams

A large range of entities falls under the general title of 'diagrams'. Hegarty, Carpenter, and Just (1991) divided them into three broad types. These are:

- *Iconic diagrams*. These are drawings or pictures that are realistic representations (usually 2D) of concrete objects, enabling comparisons to be readily made to their referents.
- *Schematic diagrams*. These are simplified and abstract representations in which aspects of the original object are depicted symbolically. Understanding such diagrams requires the mastery of the relationship of the elements of which they are composed to the original entity. Electric circuits, industrial processes, taxonomic trees, are examples from respectively physics, chemistry, and biology.
- *Graphs (or graphical diagrams)*. Each of the many sub-modes of graph shows, in various ways, the relationship between how the values of independent and dependent variables vary against each other. The relationship between the variables and the entities that they depict must be known if the meaning of a graph is to be understood.

Diagrams have a central place in the processes and outcomes of scientific enquiry. All three types are used in all the sciences (for example, in physics, chemistry, biology, geosciences) to provide representations severally at the macroscopic, sub-microscopic, and symbolic levels. This strongly suggests that how students can generate visualisations from diagrams (and vice versa) should be appreciated during modelling-based teaching. Three ways for students to learn how to generate visualisations during modelling-based teaching are:

- Developing students' awareness of why particular types of diagram are being used by a teacher at a given time during instruction. Liu, Won, and Treagust (2014) collected preliminary evidence from which such an approach could be partially developed by more extensive research and development. The use of diagrams by five secondary-school teachers of biology was recorded over a 7-month period. Three main conclusions were reached: iconic diagrams were used both to introduce and to explain topics; different types and examples of diagram were used interchangeably to support students' understanding of ideas across the three levels of representation; the use of diagrams that presented problems of understanding to students were augmented by teachers' use of (often self-generated) analogies.
- Developing students' visualisation skills by having them create diagrams. The role of student-drawn diagrams in science education has long been appreciated in respect of younger students (for instance, by Hayes, Symington, & Martin, 1994) and has gradually developed for older students. From the initial impetus provided by diSessa (2004), work has expanded to enable such diagrams to be used to

  > "record precise observations; identify the distribution of types; classify examples into categories; identify and explain key causes; integrate different ideas; contextualise the part of the whole" (Waldrip & Prain, 2012, p. 151) (and discharge many other functions).

  General principles have emerged on how the approach should be integrated into the curriculum (see, for example, Tytler, Petersen, & Prain, 2006). The key issue is that, as in all modelling-based teaching, whilst students must be freed to develop their ideas, the teacher has a central role in ensuring that major issues in respect of the representations are addressed by the students (Parnafes & Trachtenberg-Maslaton, 2014).

- Developing students' appreciation of the value of diagrams. Asking students to evaluate the use of particular diagrams can also be used to develop their awareness of the capabilities presented by different sub-types. For example, in small-scale case study of what might be done, a class of 14–15 year-old high-achieving science students in UK were asked to evaluate (for homework) the educational value of the 26 different diagrams that they had met in textbooks when studying the 'rock cycle' (M. Newberry, personal communication, 15 January 2014). The conclusions that they reached were interesting:

  - Cartoons were thought useful for younger students than themselves, but might lead to the acquiring of misconceptions.
  - Iconic diagrams, starting from simplified sketches, were the best way for students like themselves to grasp the basics of the topic.
  - A gradient of diagram-type with a steadily increasing use of labelling and arrows was found to lead to advanced understanding.

- Totally schematic diagrams were of the greatest use in revising for examinations because they summarised a great deal of information in a small space.
- Diagrams that were too complicated, pseudo-3D, mixed 2D and pseudo-3D, were all thought to be capable of being misleading.
- Whilst the students all realised that they were studying a cyclic phenomenon, it did not matter where in the cycle they started, although placing volcanoes on the left-hand side of a diagram guided them to start there.

Drawing on the emergent understanding of students has, on the basis of this case study, much to commend it.

Developing Visualisation Through the Use of Mathematical Modelling

Modelling with the use of the techniques of mathematics is an essential ingredient in all scientific enquiries beyond the elementary level. Whilst a science teacher may hope that mathematical modelling, as a technique, has already been taught to students in mathematics classes, it is very likely that the requisite knowledge and skills will have to be acquired and developed, or at least enhanced, in the science class itself. There is little principled yet practical guidance available on how this might be done. Overviews of the field of mathematical modelling and its development in general are available (for example, Lamon, Parker, & Houston, 2003), although its implications specifically for science education have been underexplored. One thing is clear: for those students who might wish to become science researchers or engineers, mastery of the techniques will be needed. Excellence in mathematical modelling is a prerequisite for advanced scientific work, the 'cutting edge' of research, for such work is at the third, or symbolic, level of representation. Alas, one consequence of the general adoption of 'science for all' seem to have been that consideration of mathematical models and modelling is usually deferred until late in the school years or indeed to university and only for those who choose to study science or engineering.

Inevitably, one approach to that teaching has been to present students with case studies of the conduct of mathematical modelling in action, for example in the explanation of the 'singing wine glass' (Voges & Joubert, 2008). The intention in this type of approach seems to be that students grasp the underlying principles by an analysis of the case study: in practice, this may, or may not, happen. An approach that is more likely to have transfer value to different problems and contexts is to explicitly teach the processes involved. Based on the work of Niss (2010) and Blum and Leiß (2007) we identify the following steps in the teaching as being to:

- Identify the major features of the situation to be mathematically modelled, such as an aspect of the world-as-experienced or a representation of an aspect of that situation in another medium (for example in concrete or diagrammatic form), and then to depict them in simplified form. To take three brief examples:

- if the problem is 'how far up a conic wineglass would represent a half-full glass?', the major feature is to identify the wineglass as being, in cross section, a triangle (Niss, 2010);
- if the problem in chemistry is 'what is camping gas?', the major feature is to identify the formula of the gas as being C_xH_y (Di Fuccia, 2013);
- if the problem is to explain the lift produced by a bird's wing, the key feature to identify is the cross-sectional shape of the wing (Paton, 1990).

- Identify the forms of mathematical representation that are relevant to the task and apply them to the task in hand. In the cases of the three examples these forms are:
  - the principles of plain geometry;
  - the general-form chemical equation for the burning of camping gas;
  - the application of the mathematical principles of aerodynamics.
- Translate the resulting mathematical model back into the language of the original problem, so that it can be seen that a suitable solution has been found.

This brief description requires a great deal of knowledge and skill on the part of a student. The student must, firstly, be able to perceive those aspects of the problem-situation that are capable of being represented mathematically; secondly, be able to select and apply the appropriate mathematical techniques; thirdly, have the character to persevere with the mathematical task in hand; fourthly, be able to translate the mathematical model back into the world-as-experienced (Niss, 2010). These prerequisites will, in themselves, have to be directly addressed and facilitated during modelling-based teaching.

Developing Visualisation Through the Use of Chemical Equations

Earlier in this chapter we summarised the complex protocols – both historical and contemporary – that are invoked in the writing of chemical equations. Given this complexity and the historical process of their evolution, which is now enjoying a plateau of agreement on form, it is not surprising that many students find great difficulty in mastering this mode of representation (see, for example, Kind, 2004). Taber (2009) has analysed the logic of the representational system and the thinking behind many of the misconceptions that students display. Although the nature, composition, and analysis of chemical equations will be a major topic in all chemistry curricula, classes based on modelling-based teaching where chemical equations may be evoked will require that the necessary skills be already firmly in place. Support for teachers in so-doing is, if needed, readily available electronically (Royal Society of Chemistry, 2014).

Developing Visualisation Through the Use of Drama

Drama does seem to have considerable potential as a way for students to learn the skills of both modelling and visualisation. Thus: the approach lends itself to the creation of novel images; it is multi-modal (in that sight, sound, and touch are all invoked); it uses anthropological analogies for ideas, with which students are comfortable, despite the anxieties of educators (Talanquer, 2007); it involves the creation and running of thought experiments; it provides an arena for extended discourse between the students and the teacher; and it entails an effective element of formative assessment as students more readily come to terms with their understandings and misunderstandings as they seek to perfect their performance in front of their peers. The principles of drama education, usually deployed in First Language teaching (see, for example, Kempe & Ashwell, 2000), can be readily invoked here, although they are only beginning to be used in science education (Dorion, 2009; McGregor, 2012).

Developing Visualisation Through the Use of Animations and Simulations

The research literature is divided over the general educational benefits of these allied modes (Eilam, 2012). However, from the analyses of that literature by Ainsworth (2008) and Mayer and Pilegard (2005), we have assembled the criteria below for the design of an animation or simulation by students that maximizes its contribution to learning through the expression of creativity, and hence to the formation of visualisations, in student modelling activities. These are that the animation or simulation:

- should only be used after the training of students in the design and interpretation of these modes;
- should be accompanied by an overview of the phenomenon in general, so that their significance – their scope and limitations – can be readily grasped;
- should be accompanied wherever possible by access to experience – direct or indirect – of the phenomenon itself;
- should make lower-order cognitive demands on the user (for instance, knowledge and application) before making any higher-order demands (for instance, comprehension and application) (Bloom, 1956);
- the overall representation should be split into parts, so that students have time and mental capacity to process the information being presented;
- should be slowed down, to reduce the cognitive load on the student. The 'slow motion' technique is an example of how this can be done for simulations (Loughran, 2014);

- should be accompanied by links to everyday life that are either included in it or in the accompanying text or commentary. This will increase the engagement of students;
- should be provided with a verbal, rather than written, accompanying text. This too will lower the cognitive load on the user;
- should include elements that encourage meta-cognition in the user, for example requirements to monitor the understanding being achieved, or requirements to evaluate the progress of it;
- should be best developed or used by a group, to facilitate the social aspects of learning.

The Need for the Development of Visualisation as an Adjunct to Modelling-Based Teaching

The student modeller, having made good use of these opportunities, will be lead to understand and be able to use the various codes of representation in order to develop visualisations of aspects of the world-as-experienced. When really successful, not only will that person possess all the knowledge that is relevant, but will also display all the relevant skills, be able to reflect on how such knowledge and skills may be used, be aware of the range of tasks to which it can be applied, together with the ways in which this can be done. That person may be said to be 'meta-visually competent' (Gilbert, 2005) as already discussed.

Many of the individual techniques for the development of the skills of visualisation that have been presented in this section have not been widely implemented, let alone orchestrated into a coherent provision within MBT. The nature of the challenge in doing is discussed in Chap. 12. However, we can address the key issue of the coordination or visualisation and modelling will imply.

Coordinating the Development of the Skills of Modelling and of Visualisation

Earlier in this chapter we identified the main skills entailed in meta-visual competence. Table 7.1 is an analysis in more detail of the abilities that a student must attain in order to display those skills.

In Chap. 4 we outlined a 'model for modelling-based teaching'. As the capabilities to model and to visualise are inter-dependent, in Table 7.2 we summarise the relationship between the mutual development/use of the two capabilities over time.

Although the relationships established in Table 7.2 seem to be theoretically consistent, they still need to be empirically analysed.

Table 7.1 Visualisation skills and the abilities associated with them

Skills	Abilities
To establish the purposes for a visualisation	1. To identify what is known about a phenomenon that is to be visualised
	2. To relate the uses of the visualisation to what is known about the phenomenon
To use the modes & sub-modes in the construction of a visualisation	(1–2) +
	3. To know the codes of representation for modes & sub-modes of representation
	4. To identify an existing visualisation that seems like that which has to be constructed
	5. To draw an analogy between the established and the nascent visualisation
	6. To select an appropriate mode or sub-mode
	7. To apply the selected mode or sub-mode to the phenomenon to be visualised
To test a visualisation	(1–7) +
	8. To run a thought experiment that includes the visualisation
	9. To establish the relationship between the behaviour of the visualisation and that of the phenomenon
To evaluate the validity of a visualisation	(1–9) +
	10. To identify the limitations of the visualisation in contrast to the behaviour of the phenomenon
	11. To relate the constructed visualisation to those for all three levels of representation of the same phenomenon
	12. To convince others of the validity of the visualisation

Table 7.2 The stages of modelling and the visualisation skills and abilities needed

Skills related to the stages of modelling	Visualisation skills and abilities needed
Creation of a proto-model:	To deal with the purposes for the visualisation in relation to the information available (1–2). To select from the codes of representation or identify an established visualisation (3–6) + 11
(A) To define the aims of the model or to understand the aims proposed by someone else	
(B) To search for information about the entity to be modelled (in the modeller's cognitive structure and/or from external sources: bibliography, empirical activities, etc.)	
(C) To select and organise the available information by having the aims of the model in mind	
(D) To use analogical reasoning or a mathematical tool in order to base the proto-model	
(E) To integrate all the previous information and experiences in the creation of a proto-model	

(continued)

Table 7.2 (continued)

Skills related to the stages of modelling	Visualisation skills and abilities needed
Expression of the proto-model → production of the artefact (model):	To create/adapt the visualisation (1–6)
(F) To visualise the proto-model	
(G) To use any of the models of representation (or a combination of them) to express the proto-model	
(H) To adapt the expressed to the mental proto-model (or vice-versa) by making comparisons between each other or between them and the aims of the model	
Test of the model:	To test the visualisation (8–10)
(I) To plan and conduct thought experiments	
(J) To plan and conduct empirical experiments	
(K) To critically analyse the results of the tests	
(L) To modify or to reject the model	
Evaluation of the model:	To evaluate the validity of the visualisation (9–11)
(M) To identify the limitations of the model by trying to use it in different contexts	
(N) To identify the scope of the model by trying to use it in different contexts	
(O) To convince others of the validity of the model	
(P) To persuade others to use a given model	

References

3D Printing. (2015). What is 3D printing? Retrieved March, 2015, from http://3dprinting.com/what-is-3d-printing/-wt

Ainsworth, S. (2008). How do animations influence learning? In D. H. Robinson & G. Schraw (Eds.), *Recent innovations in educational technology that facilitate student learning* (pp. 37–67). Charlotte, NC: Information Age Publishing.

Blackwell, A. F., & Engelhardt, Y. (1998). *A taxonomy of diagram taxonomies*. Paper presented at the thinking with diagrams 98: Is there a science of diagrams?

Bloom, B. S. (1956). *A taxonomy of educational objectives*. New York: David McKay.

Blum, W., & Leiß, D. (2007). How do students and teachers deal with modeling problems. In C. Haines, P. Galbraith, W. VBlum, & S. Khan (Eds.), *Mathematical modelling: Education, engineering and economics* (pp. 222–231). Chichester, UK: Horwood Publishing.

Brown, J. (1991). *The laboratory of the mind*. London: Routledge.

Burges, D. (1980). Teaching applications of mathematics: Mathematical modelling in science and technology. *European Journal of Science Education, 2*(4), 365–376.

Cassels, J., & Johnstone, A. (1985). *Words that matter in science*. London: Royal Society of Chemistry.

Di Fuccia, D. (2013). *Mathematical models in chemistry lessons*. Paper presented at the X Conference of the European Science Education Research Association, Nicosia, Cyprus.

diSessa, A. A. (2004). Metarepresentation: Native competence and targets for instruction. *Cognition and Instruction, 22*(3), 293–331.

Dori, Y. J., Rodrigues, S., & Schanze, S. (2013). How to promote chemistry learning through the use of ICT. In I. Eilks & A. Hofstein (Eds.), *Teaching chemistry: A studybook* (pp. 213–240). Rotterdam: Sense.

Dorion, K. (2009). Science through drama: A multiple case exploration of the characteristics of drama activities used in secondary science lessons. *International Journal of Science Education, 31*(16), 2247–2270.

Eilam, B. (2012). *Teaching, learning, and visual literacy*. Cambridge, UK: Cambridge University Press.

Ferrucci, B. J., & Carter, J. A. (2003). Technology-active mathematical modelling. *International Journal of Mathematical Education in Science and Technology, 34*(5), 663–670.

Francoeur, E. (1997). The forgotten tool: The design and use of molecular models. *Social Studies of Science, 27*, 7–40.

Gilbert, J. K. (2005). Visualization: A metacognitive skill in science and science education. In J. K. Gilbert (Ed.), *Visualization in science education* (pp. 9–27). Dordrecht, Netherlands: Springer.

Gilbert, J. K. (2008). Visualization: An emergent field of practice and enquiry in science education. In J. K. Gilbert, M. Reiner, & M. Nakhleh (Eds.), *Visualization: Theory and practice in science education* (pp. 3–24). Dordrecht, Netherlands: Springer.

Gilbert, S. W. (2011). *Models-based science teaching*. Arlington, Virginia: NSTA Press.

Gilbert, J. K., & Reiner, M. (2000). Thought experiments in science education: Potential and current realisation. *International Journal of Science Education, 22*(3), 265–283.

Gilbert, J. K., & Watts, D. M. (1983). Conceptions, misconceptions, and alternative conceptions: Changing perspectives in science education. *Studies in Science Education, 10*(1), 61–98.

Givry, D., & Roth, M.-W. (2006). Towards a new conception of conceptions: Interplay of talk, gestures and structures in the setting. *Journal of Research in Science Teaching, 14*(10), 1096–1109.

Goldin-Meadow, S. (2006). Nonverbal communication: The hand's role in talking and thinking. In W. Damon & R. M. Lerner (Eds.), *Handbook of child psychology* (6th ed., Vol. 2, pp. 336–369). Hoboken, NJ: Wiley.

Hand, B. M., Prain, V., Lawrence, C., & Yore, L. D. (1999). A writing in science framework designed to enhance scientific literacy. *International Journal of Science Education, 21*(10), 1021–1035.

Harvey, W. (1993). *On the circulation of blood and other writings*. London: Everyman Orion.

Hayes, D., Symington, D., & Martin, M. (1994). Drawing during science activity in the primary school. *International Journal of Science Education, 16*(3), 265–277.

Hegarty, M., Carpenter, P., & Just, M. (1991). Diagrams in the comprehension of scientific texts. In R. K. Barr (Ed.), *Handbook of reading research* (Vol. 2, pp. 641–668). New York: Longman.

Hegarty, M., & Waller, D. (2005). Individual differences in spatial abilities. In P. Shah & A. Miyake (Eds.), *The Cambridge handbook of visuospatial thinking* (pp. 121–169). Cambridge: Cambridge University Press.

Herrera, J. S., & Riggs, E. M. (2013). Relating gestures and speech: An analysis of students' conceptions about geological sedimentary processes. *International Journal of Science Education, 35*(12), 1979–2003.

Hodson, D. (2009). *Teaching and learning about science: Language, theories, methods, history, traditions and values*. Rotterdam, Netherlands: Sense.

IUPAC. (2014). *Compendium of chemical terminology: Gold book*. Washington, DC: IUPAC.

Johnstone, A. H. (1993). The development of chemistry teaching: A changing response to changing demand. *Journal of Chemical Education, 70*(9), 701–705.

Justi, R., & Gilbert, J. K. (2002). Modelling, teachers' views on the nature of modelling, implications for the education of modellers. *International Journal of Science Education, 24*(4), 369–387.

Kastens, K. A., Agrawal, S., & Liben, L. S. (2008). The role of gestures in geoscience teaching and learning. *Journal of Geoscience Education, 56*(4), 362–368.

Kempe, A., & Ashwell, M. (2000). *Progression in secondary drama*. Harlow, UK: Pearson.

Kind, V. (2004). *Beyond appearances: Students' misconceptions about basic chemical ideas* (2nd ed.). London: Royal Society of Chemistry.

Kosslyn, S. (2006). *Graph design for the eye and mind*. Oxford: Oxford University Press.

Kosslyn, S., Pinker, S., Smith, G., & Shwartz, S. (1982). On the demystification of mental imagery. In N. Block (Ed.), *Imagery* (pp. 131–150). Cambridge, MA: MIT Press.
Kozma, R., & Russell, J. (2005). Students becoming chemists: Developing representational competence. In J. K. Gilbert (Ed.), *Visualization in science education* (pp. 121–146). Dordrecht, Netherlands: Springer.
Kuhn, T. S. (1996). *The structure of scientific revolutions* (3rd ed.). Chicago: University of Chicago Press.
Lakoff, G., & Johnson, M. (1980). *Metaphors we live by*. Chicago: University of Chicago Press.
Lamon, S. J., Parker, W. A., & Houston, S. K. (Eds.). (2003). *Mathematical modelling: A way of life*. Chichester, UK: Horwood.
Landriscina, F. (2013). *Simulation and learning: A model-centered approach*. New York: Springer.
Levin, J. R., & Mayer, R. E. (1993). Understanding illustrations in text. In B. K. Britton, A. Woodward, & M. Binkley (Eds.), *Learning from textbooks: Theory and practice* (pp. 95–113). Hillsdale, NJ: Lawrence Erlbaum.
Levin, J. R., Shriberg, I., & Berry, J. (1983). A concrete strategy to remembering abstract prose. *American Educational Research Journal, 20*(2), 277–290.
Liu, Y., Won, M., & Treagust, D. F. (2014). Secondary biology teachers' use of different types of diagrams for different purpose. In B. Eilam & J. K. Gilbert (Eds.), *Science teachers' use of visual representations* (pp. 103–121). Dordrecht, Netherlands: Springer.
Loughran, J. (2014). Slowmation: A process of explicit visualization. In B. Eilam & J. K. Gilbert (Eds.), *Science teachers' use of visual representations* (pp. 85–102). Dordrecht, Netherlands: Springer.
Mayer, R. E., & Pilegard, C. (2005). Principles for managing essential processing in multimedia learning: Segmenting, pre-training, and modality principles. In R. E. Mayer (Ed.), *The Cambridge handbook of multimedia learning* (pp. 316–344). New York: Cambridge University Press.
McGregor, D. (2012). Dramatising science learning: Findings from a pilot study to re-invigorate elementary science pedagogy for five- to seven-year olds. *International Journal of Science Education, 34*(8), 1145–1165.
McNeill, D. (2005). *Gesture and thought*. Chicago: University of Chicago Press.
Newcombe, N. S., & Learmonth, A. E. (2005). Development of spatial competence. In P. Shah & A. Miyake (Eds.), *The Cambridge handbook of visuospatial thinking* (pp. 213–256). Cambridge: Cambridge University Press.
Newcombe, N. S., & Stieff, M. (2012). Six myths about spatial thinking. *International Journal of Science Education, 34*(6), 955–971.
Niss, M. (1987). Applications of modelling in the mathematics curriculum: State and trends. *International Journal of Mathematical Education in Science and Technology, 18*(4), 487–505.
Niss, M. (2010). Modeling a crucial aspect of students' mathematical modeling. In R. Lesh, P. L. Galbraith, C. R. Haines, & A. Hurford (Eds.), *Modeling students' mathematical modeling competencies* (pp. 43–59). New York: Springer.
Padalkar, S., & Ramadas, J. (2011). Designed and spontaneous gestures in elementary astronomy education. *International Journal of Science Education, 33*(12), 1703–1739.
Parnafes, O., & Trachtenberg-Maslaton, R. (2014). Transforming instruction: Teaching in a student-generated representations learning environment. In B. Eilam & J. K. Gilbert (Eds.), *Science teachers' use of visual representations* (pp. 271–290). Dordrecht, Netherlands: Springer.
Paton, R. (1990). Bird wings and matrices. *Journal of Biological Education, 24*(4), 273–276.
Paton, R. (1991a). An application of mathematical modelling to school biotechnology. *International Journal of Mathematical Education in Science and Technology, 22*(2), 291–296.
Paton, R. (1991b). Modelling biological processes using simple matrices. *Journal of Biological Education, 25*(1), 37–43.
Pozzer, L. L., & Roth, M.-W. (2003). Prevalence, function, and structure of photographs in high school biology textbooks. *Journal of Research in Science Teaching, 40*(10), 1089–1114.

Prain, V., & Tytler, R. (2012). Learning through constructing representations in science: A framework of representation construction affordances. *International Journal of Science Education, 34*(17), 2751–2773.

Reiner, M. (1998). Thought experiments and collaborative learning in physics. *International Journal of Science Education, 20*(9), 1043–1059.

Roth, M.-W., & Welzel, M. (2001). From activity to gestures and scientific language. *Journal of Research in Science Teaching, 38*(1), 103–136.

Rothbart, D. (Ed.). (2004). *Modeling: Gateway to the unknown: A work by Rom Harré*. Amsterdam: Elsevier.

Royal Society of Chemistry. (2014). *CPD for teachers*. Retrieved July, 2014, from http://www.rsc.org/careers/cpd/teachers

Savec, V. F., Vrtacnik, M., & Gilbert, J. K. (2005). Evaluating the educational value of molecular structure representations. In J. K. Gilbert (Ed.), *Visualization in science education* (pp. 269–297). Dordrecht, Netherlands: Springer.

Savory, C. (1976). A survey of cristal and molecular models. *Education in Chemistry, 13*, 136–141.

Scott, P., Asoko, H., & Leach, J. (2007). Students conceptions and conceptual learning in science. In S. K. Abell & N. G. Lederman (Eds.), *Handbook of research in science education* (pp. 31–56). Mahwah, N.J.: Lawrence Erlbaum.

Sorensen, R. (1992). *Thought experiments*. New York: Oxford University Press.

Swetz, F. (1989). A Historical example of mathematical modelling: The trajectory of a cannonball. *International Journal of Mathematical Education in Science and Technology, 20*(5), 731–741.

Taber, K. S. (2002). *Chemical misconceptions: Prevention, diagnosis and cure* (Vol. 1). London: Royal Society of Chemistry.

Taber, K. S. (2009). Learning at the symbolic level. In J. K. Gilbert & D. F. Treagust (Eds.), *Multiple representations in chemical education* (pp. 75–108). Dordrecht, Netherlands: Springer.

Talanquer, V. (2007). Explanation and teleology in chemical education. *International Journal of Science Education, 29*(7), 853–870.

Thomas, N. J. T. (2014). Mental imagery. In E. N. Zalta (Ed.), *The stanford encyclopedia of philosophy* (Fall 2014 ed.). Stanford: Stanford University. http://plato.stanford.edu/archives/fall2014/entries/mental-imagery/.

Tuckey, H., & Selvaratnam, M. (1993). Studies involving three-dimensional visualisation skills in chemistry: A review. *Studies in Science Education, 21*, 99–121.

Tytler, R., Petersen, S., & Prain, V. (2006). Picturing evaporation: Learning science literacy through a particle representation. *Teaching Science, 52*(1), 12–17.

Urhahne, D., Sabine, N., & Schanze, S. (2009). The effect of three-dimensional simulations on the understanding of chemical structures and their properties. *Research in Science Education, 39*(4), 495–513.

Voges, E., & Joubert, S. (2008). The singing wineglass: An exercise in mathematical modelling. *International Journal of Mathematical Education in Science and Technology, 39*(6), 725–739.

Waldrip, B., & Prain, V. (2012). Learning from and through representation in science. In B. J. Fraser, K. G. Tobin, & C. J. McRobbie (Eds.), *Second international handbook of science education* (Vol. 2, pp. 145–156). Dordrecht, Netherlands: Springer.

Wellington, J., & Osborne, J. (2001). *Language and literacy in science education*. Buckingham, UK: Open University Press.

Wilkes, K. V. (1988). *Real people: Personal identity without thought experiments*. Oxford: Oxford University Press.

Willows, D. (1978). A picture is not always worth a thousand words: Pictures as distractors in reading. *Journal of Educational Psychology, 70*(2), 255–262.

Wilson, M. (2002). Six views of embodied cognition. *Psychonomic Bulletion and Review, 9*(4), 625–636.

Winn, W. (1991). Learning from maps and diagrams. *Educational Psychology Review, 3*(3), 211–247.

Chapter 8
Analogies in Modelling-Based Teaching and Learning

Abstract The creation and use of analogies play important roles in modelling. An analogy is created when some aspects of an unknown *target* are compared with those of a *source* about which more is known. The drawing of an analogy between a target and a source involves following a series of steps, the nature of the possible relationship being bounded by a series of requirements and constraints. Suitable analogies, once created, play central roles in providing explanations for difficult models, acting as *teaching models* to promote the understanding of conceptually difficult target models. This role has been extensively investigated in science education. MBT provides an opportunity to introduce students to the creative role of analogies. A case study is presented on the creative use of analogy in each aspect of a modelling activity is facilitated by MBT. The teacher's part in the operation of both these roles – explanatory and creative – is outlined.

Figurative Language in Science Education

In order to gain new insights into the world-as-experienced, scientists and hence science educators make considerable use of figurative language. This is the use of words in such a way as to deviate from their standard definitions in order to produce novel or amended meanings for them. This usage can fall into many categories, for example metaphor, metonym, synecdoche, simile. However, the category that has proved to be of the greatest use in the world of science is analogy. In analogy, a word or phrase that has been found valuable in describing or explaining some aspect of a phenomenon about which a lot is known is used tentatively to describe or explain an aspect of a phenomenon about which less is known. It is able to do so because it is concerned with the possible relationships between them.

Analogies and Models

Since the last decades of the twentieth century, science educators and psychologists have explicitly recognised and discussed this role of analogies in science and everyday life (Duit, 1991; Harrison & Treagust, 1993; Hesse, 1966; Holyoak & Thagard,

J.K. Gilbert, R. Justi, *Modelling-based Teaching in Science Education*, Models and Modeling in Science Education 9, DOI 10.1007/978-3-319-29039-3_8

1989; Lakoff & Johnson, 1980; Vosniadou, 1989). In the literature of both areas, it seems there is an agreement upon the meaning of 'analogy', although there are distinct views about how an analogy is drawn, that is, how the likenesses identified between the 'known' and the 'unknown' are arrived at. We assume in this book that an analogy is a comparison that involves the alignment and mapping of relational structures between two domains (Gentner, 1983; Gentner & Markman, 1997). One of them – called the *base* (Gentner, 1983) or the *source* (Holyoak & Thagard, 1989) – is familiar to the enquirer, whilst the other – called the *target* (by all authors) – is partially or totally unknown.

According to Gentner, who first proposed such ideas, there are three constraints on analogical alignment and mapping. To be successful, that is, to provide new insights, an analogy should show:

- *structural consistence*, that is, it should make some isomorphic mappings between elements of the two domains (one known, one unknown) evident. This means that (i) there should only be a single one-to-one correspondence between each pair of elements, and (ii) the matching relation has to be expressed in terms of arguments. For instance, in the analogy between the atom and the solar system, the sun corresponds to the atomic nucleus, and the planets correspond to the electrons. In each such a correspondence, the elements play similar roles in a relational structure that is assumed to be common to the source and the target;
- *structural focus*, that is, it must involve common relationships across the two but need not involve common object descriptions of them. For instance, in the analogy between the atom and the solar system, it does not matter if planets have different shapes and sizes but does assume that the electrons are all of an equal size, because the mapping is concerned with their movements around a central point; and
- *systematicity*, that is, an analogy must describe systems of connected relations. In the analogy between the atom and the solar system, this is shown, for instance, by the following connected mapping relations: 'the sun and the planets attract each other' & 'the nucleus and the electrons attract each other'; and 'as the sun attracts the planets, the planets move around the sun' & 'as the nucleus attracts the electrons, the electrons move around the nucleus' (Gentner, 1983, 2002; Gentner & Markman, 1997).

The systematicity principle permits the generation of inferences. This, according to Nersessian and Chandrasekharan (2009), is something that enables analogies to be intellectually productive. It does so because it maintains higher-order relations within and between the source and the target.

Being characterised in such a way, analogies play distinct roles in science. One such role is that they are *explanatory*, that is, an analogy can be used to support the understanding of a target domain. Within this role, the use of analogies has a communicative (sometimes persuasive) purpose and/or an instructional purpose (Dunbar & Blanchette, 2001; Glynn, Britton, Semrud-Clikeman, & Muth, 1989). For instance, Maxwell used the analogical visual model he created for both communicating the knowledge he had produced about electromagnetism and as a way of trying to convince other scientists of its potential to explain allied phenomena (Nersessian, 1992). However, the main role analogies play in science is *creative*. They can help scientists to ask questions (Wilbers & Duit, 2006), to solve problems (Clement, 2008; Craig, Nersessian, & Catrambone, 2002), to think (Gentner, 2002; Nersessian, 2002), in sum, to generate knowledge. Evidence that supports this view can be found from the analysis of both the historical development of important scientific concepts, mainly those conducted by Kepler, Maxwell, Kekulé, Trautz and Lewis, Watson and Crick (Gentner, 2002; Justi & Gilbert, 2006; Nersessian, 1992), and in ethnographic studies conducted in modern laboratories (as the study reported in Dunbar, 2000; Dunbar & Blanchette, 2001). Additionally, as noted by Gentner (2002), the study conducted by Dunbar in the microbiology laboratories also shows that where many analogies are taken seriously (being discussed and extended in discussions by the scientists concerned), these prove to be highly productive. Some scientists – perhaps the most distinguished – clearly recognise both roles of analogies. According to Gentner (2002), one of the best examples is provided by Kepler, who, on many occasions, (i) stated that he used analogies and appreciate the value of doing so *("I especially love analogies, my most faithful masters, acquainted with all the secrets of nature..."* (Kepler, *The Optics*, 1904; quoted in Gentner, 2002, p. 33)), and (ii) criticised the improper uses of analogies by other scientists, for he explicitly wrote his views on how analogies may best be used in scientific thinking and communication.

Most of the authors who discuss how or why analogies play distinct roles in science associate such roles with many kinds of relationship between analogies and models. For Duit (1991), a model represents a structural identity between two domains, that is, models are outcomes of analogical reasoning in which elements of the structure of a given entity are partially mapped onto those of another (as proposed by Gentner and explained at the beginning of this section). Thus, a model represents a *general* analogical mapping between the source and the target, that is, models "may provide analogies" whilst "it is the analogical relation that makes a model a model" (Duit, 1991, p. 651). Indeed, a model may itself act as the source of an analogy. For instance, the 'water model' for the electric circuit is based on an analogical mapping between the dynamic behaviour of water flowing in a pipe and electrons "moving" in a wire. This 'water model' can itself act as a source for other models, for example of the behaviour of herds of animals during a stampede.

A similar, but more detailed view, is discussed by Clement (2008) in a book that focuses on how scientists and students construct models. For him, in the process of development of scientific explanatory models, an analogy may function as a *proto-model*, that is, "as a source analogy (source of ideas) that becomes elaborated and modified as it is incorporated into the model" (p. 271). In his GEM cycle (Fig. 2.1), there is a cyclical relationship between the actions to 'make initial observations', 'activate possible initial analogies and related model elements', and 'construct an initial model'. We incorporated these ideas in the first version of our Model of Modelling diagram (Justi & Gilbert, 2002a) (Fig. 2.2) since we recognised that analogies may be sources for models. Clement also emphasises that (i) as a model becomes more complex, several additional analogies may be drawn, that is, "an explanatory model is often constructed from multiple analogies" (p. 273); and (ii) aspects of an analogy may become part of a model "even though the analogy does not constitute the entire model" (p. 469). In other words, a model is *not in itself an analogy* and an *analogy cannot represent all the relationships between a given model and the source from which it is originated.*

Such a role for analogies in model construction has also been emphasised by Nersessian (1999). She identifies the necessary occurrence of *retrieval* in analogical problem solving, that is, of the activation of a source representation (stored in long-term memory) by the target representation (the one that comprises the representation of a problem in working memory). However, as discussed in Chap. 2, Nersessian's ideas on the relationships between analogies and models go beyond that of an analogical source domain for producing an initial model. This is so because she analyses not only the entities (analogies and models), but also the creative cognitive processes involved in modelling, which includes the use of analogical reasoning.

Such ideas are also part of our current framework for modelling. In the representations that constitute it (Fig. 2.3, 2.4, and 2.5), analogical reasoning is identified as one of the threads that constitute the string that organises the elements and processes involved in modelling. These can be depicted as forming a tetrahedron, the other threads being provided by the use of imagistic representations, the design and conduct of thought experiments, and by argumentation. This means that analogical reasoning does not only constitute the creative process by integrating different elements in the creation of a proto-model. It is also essential in the stages of:

- *expressing a model.* The selection and use of the several modes of representation to express models may be based on analogical reasoning (Aragón, Oliva-Martínez, & Navarrete, 2014; Justi & Gilbert, 2006; Oliva-Martínez & Aragón, 2009b). For instance, when a ball and stick model (using a concrete/material mode) is produced, it is assumed that the represented structure is similar to that of the substance in question. Or, when a differential equation (using a symbolic mode) is used to express the relationships between the variables attributed to a phenomenon, it is assumed that such relationships are similar to those that exist between variables in distinct comparable phenomena. For example, this occurred when the first quantitative expression for 'reaction rate' was proposed on the basis of an analogy between chemical reactions and the physical theory of heat (Laidler, 1995);
- *testing a model.* When a model is tested, the degree to which it achieves its purposes is analysed. Analogical reasoning forms the basis of the design of empirical or thought experiments because it supports the analysis of the relations mapped between the target and the source. If acceptable explanations and promising

predictions could not be identified from the relationships mapped, the model has to be modified. For instance, in the seventeenth and eighteenth centuries, it was accepted that 'affinity' (the cause of the production of new substances) was a result of corpuscles having appropriate shapes that allowed them to stick to each other. If the model of the occurrence of chemical reactions due to the affinity between particles was valid, this analogy could be extended, for instance, to explain why distinct reactions occur at different rates. As (i) predictions of rates were not found to be possible using this model, and (ii) the notions of 'shapes', 'sizes', and 'the mechanisms of particle adhesion' could not be empirically tested, the affinity model for chemical reactions had to be abandoned (Justi & Gilbert, 1999, 2006);

- *evaluating a model.* The identification of the scope and limitations of a model results from attempts to use it in different contexts. This means that evaluating a model is essentially a process of establishing and analysing comparisons between explanations and predictions provided by that model in distinctly different chemical contexts. Returning to the historical context of discussions concerning the affinity model for chemical reactions, the use of a distinct analogy resulted in a very positive evaluation of that model. This was so because Newton viewed affinity not as a spatial combination of appropriate shapes of corpuscles, but as a force acting at a distance by which bodies tended to move towards one another. If so, any chemical reaction occurring would be due to the presence of such attractive forces, which acted depending on the nature of each atom and distance between them (Mellor, 1904). At that time, it was generally accepted in Mechanics that the magnitude of a force was measured by its influence upon the motion of a body. So, if affinity between particles were interpreted from the Newtonian approach, it would be possible to determine the magnitude of this force – chemical affinity – by measuring its influence on the reactions' rate. The subsequent work of many scientists (Geoffroy, Stahl, Bergmann, etc.) improved and extended the then currently available 'affinity tables'. These showed the order in which different substances would be expected to displace one another from a given compound. Until new empirical data become available, the explanations provided by the use of the affinity tables showed that they were very helpful in predicting not only the occurrence of a series of similar reactions but also their relative rates.

The view that analogical reasoning is a vital component of modelling is supported by the converse argument that modelling is involved in the characterisation of the components of analogical reasoning. According to Gentner and Holyoak (1997), there is a consensus in the cognitive science community concerning the sub-processes, each leading towards a model, that are involved in analogical reasoning. These are:

- *Access*: one or more relevant base domains, stored in mind, is accessed or created;
- *Mapping*: a base domain is mapped onto the target domain by identifying systematic correspondences between them, and by aligning the corresponding parts of the two domains;

- *Evaluation of the match* and *Proposition of inferences*: the resulting mapping allows the individual to make deductions about the target domain, thus contributing to the construction of new knowledge;
- *Generalisation*: the extension of the inferences to all cases in which they could be applied is then possible.

Taking such characterisations into account, as well as those of the stages of modelling (Chap. 2), we may infer that the performance of these sub-processes of analogical reasoning can be supported by specific modelling stages in the following way:

- *Access* and *mapping* may occur during the stages of creating, expressing, and testing of a model.
- *Evaluation of the match* and *proposition of inferences* may mainly be performed during the testing of a model.
- *Generalisation* may be achieved as a result of the evaluation of the model.

Due to the dynamic nature of modelling, it is also possible that some of these sub-processes (specially the evaluation of the match between source and target) occur during the performance of other modelling sub-process (like 'creation' and 'expression of a model'). Such inferences are supported by an empirical study in which the elements of MBT that may have contributed to students experiencing the sub-processes of analogical reasoning were identified (Mozzer & Justi, 2011). More details on this study are provided in another section of this chapter.

Analogies in Science Teaching

From a constructivist teaching perspective – according to which learning occurs when students establish relations between what they already know and the new knowledge that they are encountering – analogies are thought to play an important role in learning. The main arguments that support this claim focus on the role of analogies in: providing visualisations of invisible or abstract entities; facilitating the understanding of abstract entities and/or relationships; increasing students' interest in the subject; making students' prior knowledge clearer to teachers; re-structuring students' prior knowledge; supporting students' creativity and reasoning (Aubusson & Fogwill, 2006; Brown, 1994; Clement, 2008; Curtis & Reigeluth, 1984; Dagher, 1994; Duit, 1991; Harrison & Treagust, 2000; Mozzer & Justi, 2012; Oliva-Martínez & Aragón, 2009b; Treagust, Duit, Joslin, & Lindauer, 1992; Vosniadou, 1989).

A good example of these roles being discharged is described by Aubusson and Fogwill (2006) using the events taking place in a series of four lessons about chemistry provided for a class of Australian year 11 (16–17 year-old) students. In the initial lesson, students performed an empirical extraction of copper from copper carbonate by initially making copper carbonate react with dilute sulphuric acid. Then, copper electrodes were placed in the resulted copper sulphate solution and copper was collected on one of the terminals. After making empirical observations of what was going on, the students were asked to model the phenomena at the sub-microscopic level by use of a 'molecule modelling' kit. The students could use their models to explain some of the empirical observations that they had made. However, when, in the next lesson, the teacher asked them to write down their explanations for the whole process, their answers, and the discussion that followed, showed that they "had not understood the ion formation of copper at all" (Aubusson & Fogwill, 2006, p. 97). Then the teacher decided to ask them to design a role-play activity to show what they thought was happening to the sub-microscopic entities involved in the processes taking place. Initially, the role-play activity revealed what the students knew and what they did not know about the existence of the copper ion in copper sulphate. This guided the teacher in taking decisions about what should be discussed in the remaining lessons. In the third and fourth lessons, there were interactions between different groups of students concerning respectively the reaction between copper carbonate and sulphuric acid and the electrolysis of the copper sulphate solution. These were excellent opportunities for the teacher "to watch students construct their ideas" (p. 99) and to modify such ideas using the explanatory power of the analogies that they created. It also became apparent to the students that the role-play approach was how they actually visualised the processes involved at the sub-microscopic level. From the detailed analysis of the lessons, Aubusson and Fogwill conclude that: "much of the learning that occurs in brought about by the discourse associated with the analogical reasoning (taking place) rather than by the role play per se" (p. 103).

As in the science context, analogies can be used in science teaching for two purposes: *the explanatory* and *the creative* (Duit & Glynn, 1996; Glynn et al., 1989). In respect of the first case, teachers and textbooks present analogies (that is, both their source and target domains, sometimes together with the relationships that have to be established between them). The hope is that, when interpreted by students in the way intended, they will contribute to a better understanding of the theme being studied. In other words, analogies are used as teaching models, presented by the teacher or the textbook to specifically support students in understanding a given aspect of a curricular model (Gilbert, Boulter, & Elmer, 2000). This approach is based on the belief that, by understanding the mapping relationships between the two compared domains, students will come to understand something about the nature or behaviour of the unknown domain.

An analysis of some studies into students' understanding of analogies (Chiu & Lin, 2005; Curtis & Reigeluth, 1984; Harrison & Treagust, 1993, 2000; Oliva-Martínez, Aragón, Mateo, & Bonat, 2001; Thiele & Treagust, 1995) shows that, on most of the occasions that analogies are used as teaching models, the authors only emphasise the similarity perspective, with discussions related to any of the constraints on the use of analogies being largely absent. For instance, Thiele and Treagust (1995) analysed 62 analogies found in Australian chemistry textbooks using criteria previously proposed by Curtis and Reigeluth (1984). Although, in many cases, the mapping relationships were made clear (26 indicated the shared attributes of source and target), no explicit discussion on how to use the analogies was provided, and only 7 analogies were accompanied by statements highlighting any unshared attributes or limitations to their use. However, many individual teachers do better than this by considering the students' background when the analogies to be used are selected. They also explicitly identified the attributes that are unshared between the source and the target. As a consequence, when limitations to the use of a given analogy are made clear, students tended to advance their understanding of concepts (Harrison & Treagust, 1993, 2000).

Additionally, many researchers have emphasised that the use of analogies in science teaching is not a simple matter, for it does not necessarily directly result in the learning intended (Dagher, 1995; Harrison & Treagust, 2006; Niebert, Marsch, & Treagust, 2012; Oliva-Martínez, Azcárate, & Navarrete, 2007; Zook, 1991). This is so because students may develop, or reinforce, misconceptions by:

- not understanding the base domain. In order to be potentially understandable, an analogy must be culturally acceptable by students, that is, they have to know the base domain, which has to be part of their cultural lives. So, teachers and textbook authors cannot take for granted, for instance, that analogies involving sports or games would be well succeeded with children or teenagers. Whilst analogies involving football tends to work with boys and girls nowadays (thanks to the popularity of this sport among both genders all over the world), analogies drawn from cricket are not likely to work with girls, mainly those from countries where this sport is not popular (or even known);
- constructing improper matches between non-comparable relations of source and target. In a book focused on the use of analogies (Harrison & Coll, 2008), many common analogies for biology, chemistry, physics, and earth and space science are detailed analysed by identifying what the authors name 'likes' (the possible mappings between the two domains), and 'unlikes' (the points where the analogy breaks down). Although improper matches can be established from both of them, the 'unlikes' tend to be the major sources of students misconceptions. For instance, one of the analogies analysed by Venville (2008) is that of a concrete model built from coloured clothespins and the structure of DNA. From manipulating that model, students can think (or reinforce their previous idea) that chemical bonds are physical links between atoms and/or molecules;
- mis-using the analogy in contexts other than the one in which it is taught. This tends to happen particularly when the matches between the comparable relations are valid only for a specific group of instances of a concept. For example, when introducing the concept of chemical bonding, some teachers and textbook authors use the magnets analogy. According to it, particles attract each other as opposite charged magnets attract each other. Although this analogy may be useful for supporting the understanding of the principle of ionic bonding (attraction between opposite charged ions), it certainly generates misunderstandings when students try to use it to explain intermolecular attractions between non-polar molecules.

In contexts in which students have to identify and/or to solve problems, the value of fostering students' using their own analogies becomes apparent, as compared to situations where they only have to understand those that are presented to them (Cosgrove, 1995; Haglund, 2013; Haglund & Jeppsson, 2012; Kaufman, Patel, & Magder, 1996; May, Hammer, & Roy, 2006; Pittman, 1999; Wong, 1993). This claim can be justified from evidence collected at all school levels, from third-grade classes (8–9 year-old children), where a student spontaneously generated an analogy to explain the causes of earthquakes and modified it in face of his colleagues counter-arguments (May et al., 2006), to university students (physics pre-service teachers in their fourth year of the teacher training programme) learning the concept of entropy (Haglund & Jeppsson, 2012). In this latter case, eight pre-service teachers worked in two groups and generated seventeen analogies. This resulted not only in them "talking science" (Lemke, 1990) during the whole lesson (something that is not common in traditional physics classrooms), but also in the group ownership of the learning that took place as manifest in the amount of time they spent trying to structurally deepen their analogies.

The drawing of analogies is a daily action of human reasoning. However, the situation is different in the learning context, when students are asked to generate analogies for a domain that is (totally or partially) unknown to them. In this case, they may neither have the repertoire of known domains from which analogies might be drawn nor recognise aspects of a given known domain as relevant for supporting a relevant relational match. This characterises a case in which they have to be creative (Kind & Kind, 2007), that is to imagine the aspect focused on the unknown domain, to search for a familiar domain that can have its structure and/or function related to it. Thus, in order to generate analogies to explain something unknown to them, students have to create a proto-model, that is, the analogy may be the starting point of a model (Brown & Clement, 1989; Clement, 2008). Therefore, one of the possible ways to support students' generation of analogies and involvement in analogical reasoning is to engage them in models and modelling-based activities.

Learning of Analogies, Models, and Modelling

As discussed in Chap. 4, we have identified five approaches to learning about models and modelling: learning curricular models, learning to use models, learning to revise models, learning to reconstruct a model, and learning to construct a model de novo (Gilbert, 2004; Justi & Gilbert, 2002b). More recently, Oliva-Martínez and Aragón (2009a, 2009b) have suggested how the development and use of analogies by students may contribute to their learning when each of the five approaches is taken. In the following paragraphs, we summarise the meaning of each of these kinds of learning, present the relationships between these two domains, and, when necessary, complement their ideas.

Very often, the *learning of curricular models*, that is, of those models assigned to each curriculum level, requires the use of teaching models. When they are properly used, that is when the base domain is really familiar to students, when the mapping between its elements and those of the target domain are clear, and when the limitations of the analogy are apparent, analogies may provide very helpful teaching models. In this case, students can learn curricular models through analogies introduced by the teacher. This characterises the explanatory use of analogies as discussed in the previous section.

Analogies can also help students to *learn to use models*. This is so because they may act as tools that allow students to use a model to interpret and make predictions about a given phenomenon (Oliva-Martínez & Aragón, 2009b). As in the previous case, here the teacher provides the model and the analogies used and may, or not, map the relation from base to the target domain. In both cases, the teacher has to pay attention to guidelines for using analogies in science teaching, like the well-known 'Teaching with Analogies' (TWA) model (Glynn, 1991; Glynn, Duit, & Thiele, 1995), according to which the mapping relations and the limitations of the analogy are made clear to students.[1]

When students *learn to revise models*, they change the models in order to use them for different purposes or in different contexts. Being able to critically analyse an analogy, thus being able to identify the features that can be involved in the drawing of fruitful relational comparisons, as well as being able to identify those that cannot be so used, are important skills needed to analyse the information gathered in the testing of models which may lead to their revision (Oliva-Martínez & Aragón, 2009b). In this case, students have enhanced opportunities to develop analogical reasoning, since they have to map the domains, to critically evaluate the match, and to propose inferences that might contribute to the revision of the original model.

As stated in Chap. 4, students may *reconstruct a model* when they recreate a model the essence of which they already know by using modes of representation other than the ones originally used to express it, and in such a creative way that some unknown details or relationships involving in the model are also learnt. Therefore, the drawing of analogies may support the reconstruction of models not only because they may allow students to express their ideas, to make decisions concerning critically analysing a given reality, and simplifying it (as emphasised by Oliva-Martínez & Aragón, 2009a). A given model may be reconstructed by being expressed through the use of a new analogy. For instance, when Mozzer and Justi (2012) asked secondary students (who had already learnt about why atoms bind to each other) to draw a comparison that supported their explanation for the nature of chemical bonding, one of them said that chemical attractions between atoms would be like family ties between a mother and her adolescent son. According to the student, a mother and her child are united to each other not physically, but from feelings, and atoms are also not physically united, but are attracted by electrostatic forces. The student also expanded on his idea of 'binding':

[1]In more details, the TWA consists of six operations that teachers have to follow: "1. Introduce target concept. 2. Cue retrieval of analog concepts. 3. Identify relevant features of target and analog. 4. Map similarities. 5. Indicate where analogy breaks down. 6. Draw conclusions." (Glynn et al., 1995, p. 261). They also admit that the order in which the operations are carried out may vary, as well as the number of times any operation is carried out.

> *"There are two attractions: from the child to the mother and from the mother to the child. Both of them are very strong but… if you were an adolescent, you would not like to arrive at a party with your mother. Your mother takes you to the party by car, but you don't want her to stay around. You don't want her to keep a tight hold on you all the time."* (Mozzer & Justi, 2012, p. 451)

and justified his analogy:

> *I would say there is an attractive force between the nucleus from one atom and the electrons from the other (atom). The nucleus from one atom attracts the electrons from the other and there is also a repulsive force between the two nuclei. But they are also close to each other.* (Mozzer & Justi, 2012, p. 451).

In this case, the student has not only expressed his model for chemical bonding using a different mode of representation, but also improved his initial analogy (whose relational correspondence involved only attraction forces) by making clear that there are also repulsive forces acting when atoms are bound to each other. Clement (1988) characterised this situation as 'analogy generation via transformation'. In terms of our approaches for learning about models and modelling, this means that the student has not only reconstructed his model, but also revised it through the use of analogical reasoning.

Finally, when students do not know a particular model beforehand, they could have the opportunity to learn how to *construct a model de novo*. Oliva-Martínez and Aragón (2009b) highlight that this occurs when students draw analogies in a creative way, that is, when they draw analogies, use them to understand a given phenomenon, and evaluate them. These authors mention some very well known studies of such a creative use of analogies (Cosgrove, 1995; Kaufman et al., 1996; Wong, 1993). However, neither these studies, nor more recent ones on students (Haglund, Jeppsson, & Andersson, 2012; Lancor, 2014; Mason, 1996; May et al., 2006; Mendonça & Justi, 2008; Spier-Dance, Mayer-Smith, Dance, & Khan, 2005), pre-service teachers (Haglund & Jeppsson, 2014; James & Scharmann, 2007), or teachers (Mozzer & Justi, 2013) generation of analogies were conducted in MBT contexts. In all these studies, the subjects were asked to generate analogies for total or partially unknown domains when learning about them (in the case of students) or when discussing about how to teach Newton's Third Law (James & Scharmann, 2007) and chemical bonding (Mozzer & Justi, 2013), in the case of teachers.

Analogies and Analogical Reasoning in MBT Contexts

In a study that was conducted in 2008, Mozzer and Justi (2012) analysed how 13–14 year-old students reason analogically when they are asked to draw analogies for a scientific target domain they are currently learning about (the nature of chemical bonding). Data were collected from individual pre- and post-teaching interviews, in which the students were asked to establish comparisons for explaining how atoms become bound to each other. This study shows that, in general, students have no difficulties in performing both the access and the mapping sub-processes.

On the other hand, they do not perform the other sub-processes (evaluation of the match, proposition of inferences, and generalization) unless they have been specifically asked to do so. As a consequence, students may either develop alternative conceptions from improper matches between non-comparable relations and/or attributes of both domains, or misuse the analogy in other contexts. For the authors, this implies that if students' drawing of analogies is to be part of teaching activities, teachers have to encourage them to perform all the analogical reasoning sub-processes. Doing so would support their involvement in those sub-processes that are self-evident to them.

Assuming that analogical reasoning is one of the three creative epistemic practices on which to base modelling (Nersessian, 2002, 2008), in a following study (Mozzer & Justi, 2011), a whole secondary class (thirty six 15–16 year-old students) was involved in modelling activities in order to learn about the process of dissolution. Such activities were planned based on the Model of Modelling framework (Justi & Gilbert, 2002a), which means that they require students' involvement in all the stages of modelling: the creation of a proto-model, the expression, testing, and evaluation of models. Mozzer and Justi intended to specifically investigate the possible relationships between students' performance of modelling activities and their capability to engage in analogical reasoning. So, in addition to the request to produce concrete models, students were specifically asked to draw analogies and to use them. Those students had studied properties of substances 8 months before, but data collected at the beginning of the study showed that none of them had an acceptable understanding about how one substance dissolves into another.

As those students had no previous experiences with the drawing of analogies in science classes, before the modelling activities started they participated in an extra activity based on the Teaching-with-Analogy (TWA) model (Glynn, 1991). In that activity, they were asked to identify the relevant characteristics of both the Bohr atomic model and the solar system, to map the similarities between the domains, with special emphasis on the relational mappings entailed, and to identify the differences between the domains. In the discussions, the teacher tried to ensure that they had understood that an analogy involves a relational comparison between two domains. The teacher had also opted not to involve them in any activity concerning the meaning of models in science because, according to her, as they had already studied atomic models, they understood that 'in science, models are not copies of realities; they rather partially represent the reality'. After the TWA activity, students participated in the modelling activities summarised in Table 8.1.

Throughout all the lessons, students worked in groups of 5 or 6. During the lessons, the teacher tried to support students' discussions by stimulating them to critically analyse, to modify, and, sometimes, to reject their analogies and models. She also asked them: to explain their choices about the modes of representation used, as well as the meaning of particular codes of representation; and to explain their ideas, to think about the coherence of the ideas expressed in their concrete models and analogies. Data were gathered from the video recording of groups and the whole class discussions. They showed that:

Table 8.1 Characterisation of modelling activities for dissolving

Activity	Brief description	Relationship with the MM diagram
1	Presentation of two systems (water+piece of chalk and water+powder grape juice) that have different behaviour in terms of the solubility of the solid material in water.	Creation of the proto-model, mainly, to have experiences with the target.
	Making of predictions about what is going to happen in each system when their components were mixed.	
	Conduction of the empirical experiments and registration of the observations.	
	Discussion about possible differences between previous predictions and actual observations.	
2	Elaboration of a model to explain the observed phenomena at the sub-microscopic level and expression of such model in a concrete mode of representation.	Creation of the proto-model: production of proto-model.
	Drawing of analogies to explain the represented phenomena.	Expression of mental model (in 3D and as an analogy).
	Identification of the similarities and differences between the aspects compared.	Testing of the models, by conducting thought experiments.
3	Observation of new behaviour of one of the systems (mix of powder grape juice in water without stirring the system).	Testing of the models, by conducting thought experiments and revising the models.
	Attempt to use the previous analogies in order to explain the new behaviour of this system.	
	If necessary, modification or rejection, followed by new elaboration, of previous analogies and/or concrete models.	
4	Extension of possible analogical inferences to other situations related to the modelled phenomenon.	Consideration of scope and limitations of the final models.

After Mozzer and Justi (2009)

- The *access* and *mapping* were mainly performed during the creation and expression of models. From the data, it was evident that the requirements of the activities (to produce a mental model, and to express it as concrete models and as analogies that could explain each observed phenomenon), and of the teacher (to explain the mapping between the compared domains) contributed to students' performance of these sub-processes of the analogical reasoning.

  For instance, in Activity 2, one of the groups said the chalk behaves like a sponge sucking up liquid. When communicating the model to the whole class, the following dialogue happened:

  Teacher: *"Why have you compared the chalk with a sponge? What are the similarities between both systems?"*

  Student 1: *"Because the sponge has small holes!"*

Student 4: *"Both of them have empty spaces, and when they are immersed in water, they liberated air."*
Student 3: *"And both of them do not react with water."*
Teacher: *"And what about the differences?"*
Student 1: *"The chalk is compact, although it is porous, it has some very small holes. The sponge has bigger holes, and more empty spaces."*

The drawing of this analogy and the explanation of its mapping relationships supported students' production and expression of a model for the behaviour of the chalk. This relationship may have occurred in the reverse order, that is, students may have had the idea that the chalk absorbs water and liberates air before they drew the analogy. In any case, the analogical sub-processes of access and mapping were performed when students created and expressed their model.

- The *evaluation of the match* and *proposition of inferences* were involved in the tests of the model. The moments where such sub-processes of analogical reasoning were identified were related to the observation of new empirical evidence (in Activity 3); to the explicit request (from Activity 3, and emphasised by the teacher) to use previous analogies to explain the new evidence; to the explicit request (from both Activities 2 and the teacher, during the discussion of Activities 2 and 3) to express the similarities and differences between the compared domains; and to the explicit request (from the teacher) to analyse their comparisons for the new knowledge built during the class discussions (after Activities 2 and 3). For instance, another group of students drew the following analogy to explain the behaviour of the juice system:

It would be like to blow dust. The particles of the juice are close to each other, like the ones of the dust. When we blow, the air separates the dust, like the water separates the particles of the juice. The differences are the size and the number of the particles in both systems. (Student 8)

After observing the mix of powder grape juice in water without stirring the system, one of her colleagues explained:

The juice went down and up, becoming spread over the water. The system is getting darker because it is going up. (Student 12)

Next, the teacher asked them to explain the new empirical observations by using their previous model. Another student answered:

The particles of the juice went up and mixed with those of the water. They continue mixed because there must be an attractive force. Otherwise, they would be at the bottom of the system. The juice would be at the bottom and the water would be above them. (Student 7)

Immediately, the student who had explained the empirical observations realised that there was an inconsistency in the relationship between the two domains:

When we blow, the dust goes up and then down, but in this system it happens the contrary: the particles of the juice go down and then they go up, mixing with the water. The dust goes up and then down due to the gravity force. Here (the juice system) *there is also gravity force, but there is attraction.* (Student 12)

After identifying this limitation of their analogy (that is, that it could explain the mixing of the particles, but not the interaction between them), the students developed a new analogy, involving magnets and things they attract:

> *A magnet attracts some things, like the water attracts the juice particles. But there are two differences. First, the magnet attracts things, but without a homogeneous distribution like the particles of juice and water. Second, the magnet and the thing that it attracts become attached to each other* (he gestured meaning physical contact), *whist the particles of the juice and those of the water become bounded, but not physically attached to each other.* (Student 8)

When doing so, it seems that immediately after expressing the analogy, the student made some deductions about the target system that resulted in the evaluation of the new match. In both cases, the student evaluated the match and proposed inferences (in any order) when the previous model/analogy was challenged.

- Finally, the *generalisation* of the analogical relationship that had been developed was established during the assessment of the model currently accepted at the beginning of Activity 4. Here students were faced with a new context in which they could try to extend their possible inferences about the process that had taken place. To do so, they had to provide evidence of the knowledge they had built up during the earlier modelling activities. For instance, in another study in which the same modelling activities were involved (Mozzer, 2013), one student explained the behaviour of the chalk and grape juice systems using the following analogy:

> *Let us consider that there are two groups of people: A and B. Some people from group A might be interested in people from group B, and vice-versa. If the ties between people in each group were very strong, people from a given group will not abandon their group to join people from the other group; they will continue in their original group. But, if the attraction between people from different groups were stronger than the relationships between people of their original group, they will leave their original group and form mixed group pairs. In the case of the particles, what makes them stay in the same group or move away from it and attract another particle are their electrical charges, that may be stronger or weaker, thus resulting in stronger or weaker attractive forces.*

Then, when he was asked to explain why oil does not dissolve in water, he answered by using the same analogy:

> *The oil does not dissolve in water because the attractive force among them, and the attractive force among the water particles, do not allow the particle of one substance to join that of the other substance. It is exactly like I said before in terms of the groups of people. There are two groups in which people are very close to each other. Then, a person from one group may, for any reason, attract the attention of some people from the other group. But those people will not abandon their group to join this person from the other group because they are very close to each other. The only limitation is the one that I had already identified before: people have wishes, and they may establish different personal relationships at the same time, whilst particles are attracted only due to electrostatic interactions.*

The facts that the student had used the same analogy and identified the same limitation show that he had evaluated the scope and limitations of his previous model and had decided that it could be used in this new context.

As evidenced in the examples discussed above, when performing these modelling activities, students drew analogies in a creative way, used them to explain the observed phenomena, and to evaluate them. As they did not know any qualitative concise model for dissolution before performing the activities, this meant they had an opportunity to learn how to *construct a model de novo* based on the drawing, expression, and revision of analogies. In that MBT context, there were specific requests (from the activity structure and from the teacher) involving the drawing, explanations, and use of analogies. This is different from other MBT learning contexts where students are asked to express their models mainly using concrete, 2D (drawings), and verbal, modes of representation. This does not imply that students in those other studies did not either express models they created through the drawing of analogies or base their models in analogies. Some examples of spontaneous drawing of analogies were observed, for instance, in the studies reported in:

- Maia and Justi (2009), when a group of students explained what they had observed in the system constituted by nitrogen dioxide (NO_2) and dinitrogen tetroxide (N_2O_4) by stating that:

  The particles collide with each other and form NO_2. Then, they continue colliding and form N_2O_4. But both processes occur simultaneously. It is like a cycle in which reactants become products and, at the same time, products become reactants;

- Mendonça and Justi (2011), when a student explained the high melting point of the NaCl by using an anthropomorphic analogy: "*The particles* (referring to the ions) *love each other. They attract each other and stay together.*"

However, as the study reported in Mozzer and Justi (2009, 2011) specifically investigated how students use analogical reasoning in modelling-based contexts, it was necessary to include the explicit request for the drawing, using, and revising of analogies in the teaching activities.

Additionally, the involvement of students in all the analogical reasoning sub-processes cannot be viewed as an autonomous process since the teacher had to guide the students in performing specific sub-processes (mainly the generalisation of the analogical relationships). Most of the time, this occurs when students have already tested their model and are confident about it. In this case, even when the analogy had been helpful in previous stages of the learning process, they tend to express their ideas by using the target (scientific) domain terms rather than those of the source (base) domain (which, in fact, is what it would be intended). This explains why examples of students using analogies spontaneously during the evaluation of their model are rare.

The results of these studies support the view that the use of MBT activities that foster students' performance of all the modelling stages can contribute to the their evolution of understanding from the establishment of mere appearance matches (those characterised by the mapping of macroscopic properties between the compared domains that, as so, have limited prediction power) (Gentner, 1983) to the drawing of actual analogies (the establishment of relational comparisons between distinct domains). This is so because, according to the literature on analogical reasoning (Vosniadou, 1989), relational comparisons are made when a more advanced knowledge about a given domain is being developed, or has just been developed; or when students are already able to integrate the macro and the sub-microscopic levels in building a more comprehensive understanding (Justi & Gilbert, 2006). Both situations were observed when those students performed those modelling activities for the teaching of dissolving. For instance, many of the students who have generated analogies involving two groups of people and the attraction between them, had previously compared the system 'juice power + water' to a generic system 'coloured ink + water', emphasising only the similarity in the descriptive characteristics of the resulted system, that is, they had produced a mere appearance match when they had just started to think about what was involved in dissolving.

Data from the studies cited above also facilitate the discussion of other contributions from the drawing of analogies in the modelling-based teaching context. One of them concerns the visualisation of sub-microscopic entities (and relationships between them) from the observation of macroscopic phenomena (Justi & Gilbert, 2006). From both the simultaneous expression of students' ideas through concrete models and analogies and the discussions concerning them, students could analyse their ideas, justify and/or rethink the codes of representation and aspects of the compared domains. In all cases, they not only used the sub-microscopic entities in a proper way to provide an explanation for the phenomena, but also improved their understanding of the meaning of such entities. For instance, at the initial activity, some students created a model in which powder juice grains were surrounded by drops of water. As they engaged in the activities, participated in the discussions orchestrated by the teacher, drew analogies in which the base domain involved the generic entity 'particle', and were questioned by the teacher about the codes of representation used in their concrete models and the meaning of particle, their model evolved so as to represent atoms and molecules. In particular, they started using the words 'atom' and 'molecule' (when discussing their models or analogies) in a way that showed a proper understanding of them (Mozzer & Justi, 2011).

It seems that teaching situations based on both modelling-based activities and the provision of explicit support by the teacher enhance both students' performance of analogical reasoning and their use of self-generated analogies. Such situations contribute to students': learning more (and more accurately) about the scientific topic; learning about analogies and models, and their role in the production of knowledge; developing visualisation skills; using and/or developing creativity and higher-order reasoning.

Concluding Remarks

The studies mentioned in this chapter show that students' ability to reason analogically, and to draw and use analogy depend on their:

- understanding of the meaning of analogies, for this is necessary in order to focus their efforts on identifying structural mapping relations;

- range of experience, for this will govern their capacity to identify suitable sources for analogies; and
- ability to recall the structure of a given source in detail, in order to design proper matches between distinct domains.

Evidence from the studies discussed in this chapter show that MBT can facilitate the development of all three.

References

Aragón, M. M., Oliva-Martínez, J. M., & Navarrete, A. (2014). Contributions of learning through analogies to the construction of secondary education pupils' verbal discourse about chemical change. *International Journal of Science Education, 36*(12), 1960–1984.

Aubusson, P. J., & Fogwill, S. (2006). Role play as analogical modelling in science. In P. J. Aubusson, A. G. Harrison, & S. M. Ritchie (Eds.), *Metaphor and analogy in science education* (pp. 93–104). Dordrecht, The Netherlands: Springer.

Brown, D. E. (1994). Facilitating conceptual change using analogies and explanatory models. *International Journal of Science Education, 16*(2), 201–214.

Brown, D. E., & Clement, J. J. (1989). Overcoming misconceptions via analogical reasoning: abstract transfer versus explanatory model construction. *Instructional Science, 18*, 237–261.

Chiu, M.-H., & Lin, J.-W. (2005). Promoting fourth graders' conceptual change of their understanding of electric current via multiple analogies. *Journal of Research in Science Teaching, 42*(4), 429–464.

Clement, J. J. (1988). Observed methods for generating analogies in scientific problem solving. *Cognitive Science, 12*, 563–586.

Clement, J. J. (2008). *Creative model construction in scientists and students - the role of imagery, analogy, and mental simulation*. Dordrecht, The Netherlands: Springer.

Cosgrove, M. (1995). A study of science-in-the-making as students generate an analogy for electricity. *International Journal of Science Education, 17*(3), 295–310.

Craig, D. L., Nersessian, N. J., & Catrambone, R. (2002). Perceptual simulation in analogical problem solving. In L. Magnani & N. J. Nersessian (Eds.), *Model-based reasoning: Science, technology, values* (pp. 167–189). New York, NY: Kluwer Academic and Plenum.

Curtis, R. V., & Reigeluth, C. M. (1984). The use of analogies in written text. *Instructional Science, 13*, 99–117.

Dagher, Z. R. (1994). Does the use of analogies contribute to conceptual change? *Science Education, 78*(6), 601–614.

Dagher, Z. R. (1995). Analysis of analogies used by science teachers. *Journal of Research in Science Teaching, 32*(3), 259–270.

Duit, R. (1991). On the role of analogies and metaphors in learning science. *Science Education, 75*(6), 649–672.

Duit, R., & Glynn, S. (1996). Mental modelling. In G. Welford, J. Osborne, & P. Scott (Eds.), *Research in science education in Europe: Current issues and themes* (pp. 166–176). London, UK: Falmer.

Dunbar, K. (2000). How scientists think in the real world: Implications for science education. *Journal of Applied Developmental Psychology, 21*(1), 49–58.

Dunbar, K., & Blanchette, I. (2001). The in vivo/in vitro approach to cognition: The case of analogy. *Trends in Cognitive Sciences, 5*(8), 334–339.

Gentner, D. (1983). Structure-mapping: A theoretical framework for analogy. *Cognitive Science, 7*(2), 155–170.

Gentner, D. (2002). Analogy in scientific discovery: The case of Johannes Kepler. In L. Magnani & N. J. Nersessian (Eds.), *Model-based reasoning: Science, technology, values* (pp. 21–39). New York, NY: Kluwer Academic and Plenum.

Gentner, D., & Holyoak, K. J. (1997). Reasoning and learning by analogy. *American Psychologist, 52*(1), 32–34.

Gentner, D., & Markman, A. B. (1997). Structure mapping in analogy and similarity. *American Psychologist, 52*(1), 45–56.

Gilbert, J. K. (2004). Models and modelling: Routes to a more authentic science education. *International Journal of Science and Mathematics Education, 2*, 115–130.

Gilbert, J. K., Boulter, C. J., & Elmer, R. (2000). Positioning models in science education and in design and technology education. In J. K. Gilbert & C. J. Boulter (Eds.), *Developing models in science education* (pp. 3–17). Dordrecht, The Netherlands: Kluwer.

Glynn, S. M. (1991). Explaining science concepts: A teaching-with-analogies model. In S. M. Glynn, R. H. Yearny, & B. K. Britton (Eds.), *The psychology of learning science* (pp. 219–240). Hillsdale, NJ: Lawrence Erlbaum.

Glynn, S. M., Britton, B. K., Semrud-Clikeman, M., & Muth, K. D. (1989). Analogical reasoning and problem solving in science textbooks. In J. A. Glover (Ed.), *Handbook of creativity*. New York, NY: Plenum Press.

Glynn, S. M., Duit, R., & Thiele, R. B. (1995). Teaching science with analogies: A strategy for constructing knowledge. In S. Glynn & R. Duit (Eds.), *Learning science in schools: Research reforming practice* (pp. 247–273). Mahwah,NJ: Lawrence Erlbaum.

Haglund, J. (2013). Collaborative and self-generated analogies in science education. *Studies in Science Education, 49*(1), 35–68.

Haglund, J., & Jeppsson, F. (2012). Using self-generated analogies in teaching of thermodynamics. *Journal of Research in Science Teaching, 49*(7), 898–921.

Haglund, J., & Jeppsson, F. (2014). Confronting conceptual challenges in thermodynamics by use of self-generated analogies. *Science & Education, 23*(7), 1505–1529.

Haglund, J., Jeppsson, F., & Andersson, J. (2012). Young children's analogical reasoning in science domains. *Science Education, 96*(4), 725–756.

Harrison, A. G., & Coll, R. K. (Eds.). (2008). *Using analogies in middle and secondary science classrooms*. Thousand Oaks, CA: Corwin Press.

Harrison, A. G., & Treagust, D. F. (1993). Teaching with analogies: A case study in grade-10 optics. *Journal of Research in Science Teaching, 30*(10), 1291–1307.

Harrison, A. G., & Treagust, D. F. (2000). Learning about atoms, molecules, and chemical bonds: A case study of multiple-model use in grade 11 chemistry. *Science Education, 84*(3), 352–381.

Harrison, A. G., & Treagust, D. F. (2006). Teaching and learning with analogies: Friend or foe? In P. Aubusson, A. Harrison, & S. M. Ritchie (Eds.), *Metaphor and analogy in science education* (pp. 11–24). Dordrecht, The Netherlands: Springer.

Hesse, M. (1966). *Models and analogies in science*. Notre Dame, IN: Notre Dame Press.

Holyoak, K. J., & Thagard, P. (1989). Analogical mapping by constraint satisfaction. *Cognitive Science, 13*(3), 295–355.

James, M. C., & Scharmann, L. C. (2007). Using analogies to improve the teaching performance of preservice teachers. *Journal of Research in Science Teaching, 44*(4), 565–585.

Justi, R., & Gilbert, J. K. (1999). History and philosophy of science through models: The case of chemical kinetics. *Science & Education, 8*(3), 287–307.

Justi, R., & Gilbert, J. K. (2002a). Modelling, teachers' views on the nature of modelling, implications for the education of modellers. *International Journal of Science Education, 24*(4), 369–387.

Justi, R., & Gilbert, J. K. (2002b). Science teachers' knowledge about and attitudes towards the use of models and modelling in learning science. *International Journal of Science Education, 24*(12), 1273–1292.

Justi, R., & Gilbert, J. K. (2006). The role of analog models in the understanding of the nature of models in chemistry. In P. J. Aubusson, A. G. Harrison, & S. M. Ritchie (Eds.), *Metaphor and analogy in science education* (pp. 119–130). Dordrecht, The Netherlands: Springer.

Kaufman, D. R., Patel, V. L., & Magder, S. A. (1996). The explanatory role of spontaneously generated analogies in reasoning about physiological concepts. *International Journal of Science Education, 18*(3), 369–386.

Kind, P. M., & Kind, V. (2007). Creativity in science education: Perspectives and challenges for developing school science. *Studies in Science Education, 43*(1), 1–37.

Laidler, K. J. (1995). *The world of physical chemistry* (2nd ed.). Oxford, UK: Oxford University Press.

Lakoff, G., & Johnson, M. (1980). *Metaphors we live by*. Chicago, IL: University of Chicago Press.

Lancor, R. A. (2014). Using student-generated analogies to investigate conceptions of energy: A multidisciplinary study. *International Journal of Science Education, 36*(1), 1–23.

Lemke, J. L. (1990). *Talking science: Language, learning and values*. Norwood, NJ: Ablex.

Maia, P. F., & Justi, R. (2009). Learning of chemical equilibrium through modelling-based teaching. *International Journal of Science Education, 31*(5), 603–630.

Mason, L. (1996). Collaborative reasoning on self-generated analogies: Conceptual growth in understanding scientific phenomena. *Educational Research and Evaluation, 2*(4), 309–350.

May, D. B., Hammer, D., & Roy, P. (2006). Children's analogical reasoning in a third-grade science discussion. *Science Education, 90*(2), 316–329.

Mellor, J. W. (1904). *Chemical statics and dynamics*. London, UK: Longmans Green.

Mendonça, P. C. C., & Justi, R. (2008). Usando Analogias com Função Criativa: Uma nova estratégia para o Ensino de Química (Using analogies with creative function: A new strategy for chemistry teaching). *Educació Química, 1*, 24–29.

Mendonça, P. C. C., & Justi, R. (2011). Contributions of the Model of Modelling diagram to the learning of ionic bonding: Analysis of a case study. *Research in Science Education, 41*(4), 479–503.

Mozzer, N. B. (2013). *O Entendimento Conceitual do Processo de Dissolução a partir da Elaboração de Modelos e sob a Perspectiva da Teoria de Campos Conceituais [Students' conceptual understanding of dissolving in a modelling-based context and from the perspective of the theory of conceptual fields]*. PhD thesis, Universidade Federal de Minas Gerais, Brazil.

Mozzer, N. B., & Justi, R. (2009). *Introdução ao Tema Dissolução através da Elaboração de Analogias pelos Alunos Fundamentada na Modelagem [Introduction to the topic dissolving from students' drawing of analogies in a modelling-based teaching context]*. Paper presented at the VII Encontro Nacional de Pesquisa em Educação em Ciências [VII Brazilian Conference on Research in Science Education], Florianópolis.

Mozzer, N. B., & Justi, R. (2011). Students' analogical reasoning when participating in modelling-based teaching activities. In C. Bruguière, A. Tiberghien, & P. Clément (Eds.), *EBook proceedings of the ESERA 2011 conference - science learning and citizenship* (pp. 764–769). Lyon, France: Université Lyon.

Mozzer, N. B., & Justi, R. (2012). Students' pre- and post-teaching analogical reasoning when they draw their analogies. *International Journal of Science Education, 34*(3), 429–458.

Mozzer, N. B., & Justi, R. (2013). Science teachers' analogical reasoning. *Research in Science Education, 43*(4), 1689–1713.

Nersessian, N. J. (1992). How do scientists think? Capturing the dynamics of conceptual change in science. In R. N. Giere (Ed.), *Cognitive models of science* (pp. 3–44). Minneapolis, MN: University of Minnesota Press.

Nersessian, N. J. (1999). Model-based reasoning in conceptual change. In L. Magnani, N. J. Nersessian, & P. Thagard (Eds.), *Model-based reasoning in scientific discovery* (pp. 5–22). New York, NY: Kluwer and Plenum.

Nersessian, N. J. (2002). The cognitive basis of model-based reasoning in science. In P. Carruthers, S. Stich, & M. Siegal (Eds.), *The cognitive basis of science* (pp. 133–153). Cambridge, UK: Cambridge Univesity Press.

Nersessian, N. J. (2008). *Creating scientific concepts*. Cambridge, MA: MIT Press.

Nersessian, N. J., & Chandrasekharan, S. (2009). Hybrid analogies in conceptual innovation in science. *Cognitive Systems Research, 10*(3), 178–188.

Niebert, K., Marsch, S., & Treagust, D. F. (2012). Understanding needs embodiment: A theory-guided reanalysis of the role of metaphors and analogies in understanding science. *Science Education, 95*(5), 849–877.

Oliva-Martínez, J. M., & Aragón, M. M. (2009a). Aportaciones de las analogías al desarrollo de pensamiento modelizador de los alumnos en química [Contribution of analogies to develop modeling thought of chemistry students]. *Educacion Quimica, 20*(1), 41–54.

Oliva-Martínez, J. M., & Aragón, M. M. (2009b). Contribución del aprendizaje con analogías al pensamiento modelizador de los alumnos en Ciencias: Marco Teórico [Contribution of learning with analogies to the modeling thought of science students]. *Enseñanza de las Ciencias, 27*(2), 195–208.

Oliva-Martínez, J. M., Aragón, M. M., Mateo, J., & Bonat, M. (2001). Una propuesta didáctica basada en la investigación para el uso de analogías en la enseñanza de las ciencias [A teaching proposal based on the investigation about the use of analogies in science education]. *Enseñanza de las Ciencias, 19*(3), 453–470.

Oliva-Martínez, J. M., Azcárate, P., & Navarrete, A. (2007). Teaching models in the use of analogies as a resource in the science classroom. *International Journal of Science Education, 29*(1), 45–66.

Pittman, K. M. (1999). Student-generated analogies: Another way of knowing. *Journal of Research in Science Teaching, 36*(1), 1–22.

Spier-Dance, L., Mayer-Smith, J., Dance, N., & Khan, S. (2005). The role of student-generated analogies in promoting conceptual understanding for undergraduate chemistry students. *Research in Science and Technological Education, 23*(2), 163–178.

Thiele, R. B., & Treagust, D. F. (1995). Analogies in chemistry textbooks. *International Journal of Science Education, 17*(6), 783–795.

Treagust, D. F., Duit, R., Joslin, P., & Lindauer, I. (1992). Science teachers' use of analogies: Observations from classroom practice. *International Journal of Science Education, 14*(4), 413–422.

Venville, G. J. (2008). Effective biology analogies. In A. G. Harrison & R. K. Coll (Eds.), *Using analogies in middle and secondary science classrooms* (pp. 82–126). Thousand Oaks, CA: Corwin Press.

Vosniadou, S. (1989). Analogical reasoning as a mechanism in knowledge acquisition: A developmental perspective. In S. Vosniadou & A. Ortony (Eds.), *Similarity and analogical reasoning* (pp. 413–437). Cambridge, UK: Cambridge University Press.

Wilbers, J., & Duit, R. (2006). Post-festum and heuristic analogies. In P. J. Aubusson, A. G. Harrison, & S. M. Ritchie (Eds.), *Metaphor and analogy in science education* (pp. 37–49). Dordrecht, The Netherlands: Springer.

Wong, E. D. (1993). Understanding the generative capacity of analogies as a tool for explanation. *Journal of Research in Science Teaching, 30*(10), 1259–1272.

Zook, K. B. (1991). Effects of analogical processes on learning and misrepresentation. *Educational Psychology Review, 3*(1), 41–72.

Chapter 9
Learning About Science Through Modelling-Based Teaching

Abstract MBT provides an effective way to learn about the nature of the scientific enterprise, a major aim of contemporary science education. The established view of the nature of science, the consensus view that asserts that the structure of the enterprise of science is identical whenever it is conducted, has been subject to extensive criticism. In order to investigate the contribution of MBT to an understanding of the scientific enterprise, the 'Family Resemblance' and the 'Whole Science' views of it were adopted. By doing so, we were able to investigate the possible contribution of each aspect of the activity of modelling to an understanding of the enterprise of science within a broader view. It was found that MBT did so, but that the development of understanding was only consolidated when several modelling activities had been undertaken. Finally, a case study of the contribution of the teacher to the development of this understanding is presented.

Initial Comments

Over recent decades a consensus has emerged among science educators, science teachers, and parts of the general public, which accepts that science is an important human enterprise that cannot be safely ignored. Thus, all citizens should have some knowledge of and about science (Abd-El-Khalick, Bell, & Lederman, 1998; Allchin, 2013; Department for Education, 2014; Driver, Leach, Millar, & Scott, 1996; Millar & Osborne, 1998; National Research Council, 2012). The part of such knowledge that is not related to specific scientific content has been named *nature of science* (NOS). The progressive recognition of its relevance can be evidenced by the huge number of publications both about the meaning of the phrase 'nature of science' (Abd-El-Khalick, 2012; Allchin, 2011, 2012b; Alters, 1997a, 1997b; Chalmers, 1982; Erduran & Dagher, 2014; Hodson, 2014b; Lederman, 2006; Matthews, 2012; Schwartz, Lederman, & Abd-El-Khalick, 2012; Smith, Lederman, Bell, McComas, & Clough, 1997) and about how to include it (or some of its elements) into science education (Abd-El-Khalick, 2013; Akerson & Hanuscin, 2007; Allchin, 2013; Allchin, Andersen, & Nielsen, 2014; Donnelly, 2001; Duschl & Grandy, 2013; Eastwood et al., 2012; Matthews, 1998; Osborne, Collins, Ratcliffe, Millar, & Duschl, 2003; Rudge & Howe, 2009; Schwartz & Crawford, 2006; Schwartz, Lederman, & Crawford, 2004; Tolvanen, Jansson, Vesterinen, & Aksela, 2014).

J.K. Gilbert, R. Justi, *Modelling-based Teaching in Science Education*, Models and Modeling in Science Education 9, DOI 10.1007/978-3-319-29039-3_9

These issues are analysed in this chapter in order to provide a framework for the discussion of knowledge about science in MBT contexts.

Nature of Science

Since the shift from only teaching scientific content to also addressing nature of science has attracted the attention of science educators, there have been attempts to identify the meaning, the scope, and the constituent elements of NOS. It is beyond the scope of this chapter to review all relevant proposals (which may be found, for instance, in an excellent recent review (Hodson, 2014b)). However, the aim of the chapter requires the presentation of the characteristics of some major approaches.

By taking a series of perspectives and methodological approaches, some groups of science educators have viewed NOS as 'lists of tenets'. The most well-known of them is that of Lederman and his colleagues, for whom NOS concerns the values and epistemological assumptions underlying scientific knowledge and its development (Abd-El-Khalick et al., 1998). By using three criteria (the accessibility of the knowledge of such aspects of NOS to students, the general consensus about the nature of such aspects of NOS, and the value of understanding the aspects of NOS for all citizens) (Lederman, 2006), they describe NOS in terms of the following list of tenets:

- scientific knowledge is tentative, empirically based, subjective (theory-laden), partly the product of human inference, imagination, and creativity, socially and culturally embedded;
- observations are different from inferences; and
- scientific theories and laws are different from each other and have distinct functions.

Such ideas have lead to a series of empirical studies conducted over the last two decades or so, mainly aiming at characterising students' and teachers' view on NOS, as well as at proposing and validating an adequate instrument for use in such characterizations. Many conference and journal papers have been published based on such studies. This has resulted in the propagation of this view of NOS, together with the use of the related instruments, throughout the world (for instance, Cakici & Bayir, 2012; Deng, Chen, Tsai, & Chai, 2011). These tenets of NOS have been emphasised in official documents, as shown by McComas and Olson (1998). Such research and policy level endorsements of this NOS list have contributed to this view having attained some degree of 'consensus' within the science education community.

However, it seems that, in fact, there is no strong consensus about the attainment of and the authority of such a view about the nature of science. For instance, Osborne et al. (2003) affirm that the 'consensus' view is a simplified view of the nature of science. These authors conducted a three-stage Delphi study with 23 experts (science educators, scientists, historians, philosophers, and sociologists of science, science communicators, and science teachers) to see whether there was a consensus agreement among such an expert community about what should be taught to students about the nature of science. The final outcome was a set of nine ideas about science considered to be essential for the school science curriculum:

- six of them are related to methodology of science: scientific methods and critical testing; creativity; science and questioning; diversity in scientific thinking; the analysis and interpretation of data; and hypothesis and prediction;
- two are related to the nature of scientific knowledge: the historical development of scientific knowledge; and science and certainty; and
- one is related to the institutions and social practices of science: cooperation and collaboration in the development of scientific knowledge.

The results obtained in this empirical study – involving a broad community with interest or engagement in science and science education – emphasise the importance of introducing discussion about aspects concerning the methods of scientific enquiry into school curricula. This is different from a separation between scientific process and nature of science that was proposed by Lederman's group.

Assuming that nature of science is a hybrid domain comprised of elements from history, sociology, and philosophy of science, and cognitive science, McComas (2008) lists core NOS ideas, derived from these domains, to be introduced to students. According to him, his list corroborates the previous 'consensus' views of NOS. However, some items on the list had not been previously emphasised, for example,

> science has developed through 'normal science' and 'revolution' as described by Thomas Kuhn; science and technology impact on each other, but they are not the same; and science and its methods cannot answer all questions (p. 251),

whilst others are presented in a more detailed way, for example,

> knowledge production in science shares habits of mind, norms, logical thinking and methods such as careful observation and data recording, truthfulness in reporting, etc. (p. 251).

The use of such lists has been criticised on the grounds that:

- The separation of 'scientific practices' (like those involved in the generation, testing, definition of validity and reliability, and communication of scientific knowledge) from 'nature of science' is artificial, since the former are a key part of science (Allchin, 2011; Grandy & Duschl, 2007; Irzik & Nola, 2011; Ryder, Leach, & Driver, 1999; van Dick, 2011).
- The NOS lists do not take into account the existence of distinct scientific disciplines (biology, chemistry, geology, physics, etc.) each requiring specific interpretations of the tenets. So, such lists cannot characterise science as being

homogeneous, and they cannot be fixed and timeless (Elby & Hammer, 2001; Irzik & Nola, 2011; Ryder et al., 1999; van Dick, 2011; Wong & Hodson, 2009).
- Although the tenets from all such lists are applicable to science, they are not specific features of science, that is, they can also characterise many other forms of human practice (Irzik & Nola, 2014; van Dick, 2011).
- Science cannot be characterised by a single list of tenets. To do so is to reinforce common stereotypes (Allchin, 2011; Eflin, Glennan, & Reisch, 1999; Wong & Hodson, 2009), and insufficiently capturing the contextual features of science (Matthews, 2012).
- The image of science portrayed overall by NOS lists – that, for instance, it does not mention any aims for scientific enquiry – is narrow (Irzik & Nola, 2011; van Dick, 2011).
- Lists tend to be used in science classrooms as NOS checklists, thus providing students and teachers with little opportunity to analyse situations and to form their own views (Matthews, 2012) and/or to be able to apply such a declarative understanding in ordinary contexts (Allchin, 2013; van Dick, 2011).

Criticisms like these have supported the generation of alternative proposals for characterising NOS. We highlight the two of these that, in our perception, are currently received much attention: the 'Family Resemblance' and the 'Whole Science' views.

The *family resemblance approach* was proposed by Irzik and Nola (2011) based on Wittgenstein's recognition that "not all terms can be defined in terms of necessary and sufficient conditions or by specifying essences or natures" (Irzik & Nola, 2014, p. 1010). According to them, "members of a family can each resemble one another in some respects but not in others" (Irzik & Nola, 2011, p. 594). By analogy, each of the scientific disciplines would have similar and dissimilar characteristics. They also highlight that, although some characteristics (like observing, and inference) are common to all sciences, they can be used neither to define science nor to demarcate it from other types of knowledge. So, from the idea of family resemblance, rather than identifying universal characteristics, several features are grouped into eight categories organized into two dimensions: *science as a cognitive-epistemic system of thought and practice* and *science as a social-institution system*. These can be organized as in Table 9.1 (for more details on each of the features and categories, see Irzik and Nola (2014) and Erduran and Dagher (2014)).

Irzik and Nola admit that the eight categories and the features they included in each of them are not exhaustive in order to "capture the structural feature of NOS in a systematic and comprehensive way" (Irzik & Nola, 2014, p. 1009). Additionally, as they are not fixed and can be develop historically, their use can characterise the nature of all sciences, showing their similarities and particularities – although not being sufficient to define science and to demarcate it from other human enterprises (since, for instance, observations and predictions are made in other areas).

In their most recent publication, Irzik and Nola (2014) recognise that their approach is a theoretical one, and express their wish that it could be pedagogically

Table 9.1 Elements of the family resemblance approach

Dimensions	Categories	Included features
Science as cognitive-epistemic system of thought and practice	Process of enquiry	Posing questions; making observations; collecting and classifying data; formulating hypotheses; constructing theories and models; comparing alternative theories and models.
	Aims and values	Prediction; explanation; consistency; simplicity; fruitfulness; viability; high confirmation; testability; empirical adequacy.
	Methods and methodological rules	A series of systematic methods and rules that scientists use to produce reliable knowledge.
	Scientific knowledge	The outcome of scientific methods and methodological rules expressed in laws, theories, models, and collections of observational reports and experimental data.
Science as a social-institution system	Professional activities	Activities performed by scientists in order to communicate and certify knowledge (attending academic meetings, presenting and publishing their ideas, reviewing manuscripts and grants proposals, writing research projects and seeking founds for them), and serving certain social goals (doing consulting work, and informing the public about matters of interest). Thus, they are expected to follow social and ethical norms.
	Scientific ethos	The norms of conduct in science like: intellectual honesty; respect for research subjects and the environment; disinterestedness; openness to free and critical discussions, and to share ideas, data, etc.
	Social certification and dissemination of scientific knowledge	The process of peer review; the publishing of papers, chapters, and books; the acceptance of the scientist's ideas by the scientific community; the use of such ideas; and their inclusion into school curricula.
	Social values	Freedom; respect for the environment; social utility of the scientific knowledge.

After Irzik & Nola, 2014

useful. Thus, they provide only some general suggestions on how to use it in science education. Other more detailed suggestions are offered by Erduran and Dagher (2014), but also without empirical support.

Thinking in the educational context, Douglas Allchin endorses the view that the understanding of nature of science has to be *functional*. So, he proposes a profile (named *Dimensions of Reliability in Science*) that is comprised of elements of nature of science that could be used to assess nature of science knowledge. This would enable the analysis of any situation involving science to take place, thus contributing to the science literacy of individuals (Allchin, 2011). Table 9.2 shows the main dimensions.

Table 9.2 Partial inventory of dimensions of reliability in science

Epistemic categories	Dimensions	Possible items
Observational	Observations and measurements	Accuracy; role of systematic study (versus anecdote); completeness of evidence; robustness (agreement among different types of data).
	Experiments	Controlled experiment; blind and double blind studies; statistical analysis of error; replication and sample size.
	Instruments	New instruments and their validation; models and model organisms; ethics of experimentation.
Conceptual	Patterns of reasoning	Evidential relevance; verifiable information versus value; role of probability in inference; alternative explanations; correlations versus causation.
	Historical dimensions	Consilience with establish evidence; role of analogy, interdisciplinary thinking; conceptual change; error and uncertainty; role of imagination and creative syntheses.
	Human dimensions	Spectrum of motivations for doing science; spectrum of human personalities; confirmation bias, role of prior beliefs; emotional versus evidence-based perceptions of risk.
Sociocultural	Institutions	Collaboration and competition among scientists; forms of persuasion; credibility; peer review and response criticism; resolving disagreement; academic freedom.
	Biases	Role of cultural beliefs (ideology, religion, nationality, etc.); role of gender bias; role of racial and class bias.
	Economics/funding	Sources of funding; personal conflict of interest.
	Communication	Norms for handling scientific data; nature of graphs; credibility of various scientific journals and news media; fraud or other forms of misconduct; social responsibility of scientists.

Allchin, 2013, p. 24

As Allchin views the profile as including all dimensions of reliability in scientific practice, he also suggests calling it *Whole Science*. This framework can be used to characterise students' learning about science "*by the degree, both in breadth and depth, to which a student is informed about the factors that shape the reliability of scientific claims*" (Allchin, 2011, p. 528, original emphasis). This would be expressed when the student applies his/her knowledge to analyse a given situation, that is, from a well-informed analysis, rather than being from answering a question about a specific NOS item.

Nature of Science and Science Education

How science education is conducted depends directly on the aims attributed to it by those who participate in the teaching and learning processes. Such aims are, explicitly or implicitly, expressed in policy and research publications in the area. For instance, the NRC report 'Taking Science to Schools' (Duschl, Schweingruber, & Shouse, 2007) states that:

> Science is both a body of knowledge that represents current understanding of natural systems and the process whereby that body of knowledge has been established and is being continually extended, refined, and revised. Both elements are essential: one cannot make progress in science without an understanding of both. Likewise, in learning science one must come to understand both the body of knowledge and the process by which this knowledge is established, extended, refined, and revised (p. 26)

In a recently published position paper, Hodson (2014a) extends such ideas by discussing four learning goals that, even with possible small differences in nomenclature, are widespreadly accepted nowadays: *learning science*, *learning about science*, *doing science* and *learning to address socio-scientific issues*. In brief,

> - *Learning science* (refers to) acquiring and developing conceptual and theoretical knowledge.
> - *Learning about science* (implies) developing an understanding of the characteristics of scientific inquiry, the role and status of the knowledge it generates, the social and intellectual circumstances surrounding the origin and development of important scientific theories, the ways in which the scientific community establishes and monitors professional practice, including robust knowledge of the linguistic conventions for reporting, defending, scrutinizing and validating scientific claims, and awareness of the complex interactions among science, technology, society and environment.
> - *Doing science* (means) engaging in and developing expertise in scientific inquiry and problem-solving.
> - *Addressing socio-scientific issues* (SSIs) (denotes) developing the critical skills to confront the personal, social, economic, environmental and moral-ethical aspects of SSIs. (Hodson, 2014a, p. 2537)

Taken together, the achievement of these four goals may contribute to a comprehensive and authentic learning of science – which justifies all efforts to promote them in classes worldwide.

Having recognised the value of such goals, the consideration of the focus of this chapter turns to 'learning about science'. This specific form of learning has often been understood as meaning 'learning knowledge of nature of science'. However, we do not think this association is always possible. Assuming Hodson's characterisation, learning about science can only be associated with learning of NOS *if* NOS is viewed from a broad perspective, that is, not from list-based approaches that, as previously discussed, have limitations as ways of characterising science. For instance, the Lederman list, which does not include issues of scientific enquiry among its tenets (thus separating the process and products of science), also does not include many elements that characterise learning about science. The NOS teaching from its perspective also does not seem to satisfy what is expected by the most recent USA Framework for K-12 Science Education:

> The learning experiences provided for students should engage them with fundamental questions about the world and with how scientists have investigated and found answers to those questions. (National Research Council, 2012, p. 9)

On the other hand, those views of NOS that include cognitive, epistemic, and social dimensions of science (like the 'Family Resemblance' and the 'Whole Science' views) are totally consistent with elements that characterise learning about science. Those who are considered to be scientific literate citizens from these perspectives are likely to be able: to assess the reliability of scientific claims; to engage in public discussion on socio-scientific issues (including being able to pose their own questions); to view science as a complex and challenging enterprise; to become more creative and critical thinkers.

Assuming, again, the comprehensiveness of Hodson's characterisation of learning about science, as well as the ample spectrum of dimensions of science that constitute the 'Family Resemblance' and the 'Whole Science' approaches, we move on to discuss how we view the possible contributions of MBT to learning about science.

MBT as a Way to Support Learning About Science

At the time that this book is been writing (2012–15), we are convinced that MBT can contribute to students' understanding of both the epistemic basis of science (that is, understanding of how we know what we know) (Duschl & Grandy, 2013), and of the cognitive and social dimensions of science (as characterised in the 'Family Resemblance' and 'Whole Science' approaches). The justifications for this claim are discussed next. In so doing, we are not saying that this modelling approach is exclusive in contributing to students' learning about science. Our arguments also apply, to a lesser or greater extent, to other MBT approaches. They must also share the basic principles of the Model of Modelling approach, that is, they must actually engage students in modelling practices in order to help them construct knowledge by producing a model *de novo*.

Contributions from Engagement in the Stages of Modelling

As emphasised in previous chapters, each of the stages of modelling involves specific actions that require specific skills to be performed. This results in MBT being a rich and complex approach in terms of students performing a scientific practice (in fact, many practices – argumentation, visualisation, conduct of enquiry, drawing of explanations, etc. – that are closely related to modelling), which has been recognised in the literature as a way to support students understanding of what science is about (Duschl & Grandy, 2013; Hodson, 2009; Osborne, 2014).

In general, the participation in all modelling stages can make students gain insight into how scientists generate knowledge from modelling and using models as cognitive artefacts that can support creative thinking. This is so mainly because the knowledge that students construct during MBT activities is built from their previous knowledge and experiences, by using the available resources and their creativity, is new for them, and tends to be more useful than their previous knowledge. Even assuming that (i) scientists have special interests in producing knowledge; (ii) their background is multifaceted and deeper than that of the students; and (iii) the processes of disseminating and validating the products of their work engaged in by scientists are more complex and are sustained for much longer periods than those experienced by students, that experience is essential in the support of students' understanding of modelling as a scientific practice.

The establishment of the aims of a model to be developed, or the understanding of the aims for it that are proposed by someone else, followed by the search for information about the entity to be modelled, were identified in Chap. 4 (Table 4.1) as the starting points of the creation of a proto-model. These activities are also features of the process of scientific enquiry as identified in the broad NOS approaches discussed in this chapter. They can be viewed as evidence that scientific knowledge is always produced for a purpose: to contribute to the solution of a problem, or as a specific part of a broader theoretical or empirical enterprise. Additionally, information that is acquired – something that is identified as the aims of the model became clearer and as initial information is gathered – can be obtained by students from sources such as:

- their own cognitive structure and available bibliography. This shows that science is an on-going process of building a network in which apparent divergent items can be connected, that scientific knowledge is not just created 'from nothing' (as if by magic);
- empirical activities. In these, knowledge is acquired for a distinct purpose, not just as an accumulation, as is the case in traditional teaching.

The selection and organisation of available information, together with the use of analogical reasoning or mathematical tools, may provide excellent opportunities to enculturate students into the creative practices of science. The performance of these sub-stages may, at the same time, result in the development of students' skill in analogical reasoning and in the formulation of hypotheses, as well as in a recognition of the role of mathematics in science – practices that can be shown as essential in the historical development of many topics that constitute the school curriculum.

The integration of the then current available information and experiences into the creation of a proto-model may be a complex, but highly significant, experience for students understanding of the processes of scientific enquiry. Additionally, as discussed in Chap. 6, the stage of creating a proto-model also requires that students deal with evidence and produce arguments to support their models (or their choices during the creation of that model). This may also become an opportunity for students to appreciate the tentativeness of the scientific knowledge. Such a development results from their recognition that (i) evidence may be incomplete (that is, it only partially supports a justification), and/or (ii) a given item of evidence may be interpreted from distinct theoretical backgrounds, thus leading to completely different arguments. Particularly in the cases where the entity to be modelled is an abstract one (frequently so in chemistry teaching), the creation of the proto-model also requires that students deal with the purposes of visualisation – another essential scientific practice that may be part of modelling (mainly when it involves abstract topics) (as discussed in Chap. 7).

Visualisation and argumentative practices are also involved in the expression of the proto-model. However, perhaps the clearest contribution of the performance of this stage to the enhancement of students learning about science is related to the necessary use of many forms of representation of science. By using concrete materials, computer, drawings, gestures, words, etc., students can:

- realise the need of using several modes of scientific representation;
- understand the possible existence of multiple ways to represent a model;
- develop their ability to identify the adequacy of each of these modes of representation depending on their purposes at particular times or on their audience;
- develop their ability to translate one mode of representation into another;
- understand the possible roles and power of specific modes of representation in terms of expressing ideas and making them more convincible.

Additionally, as the act of expressing a given idea may entail reflection about it, students may also realise the interplay between creating and expressing a model, which is another evidence of the non-linearity and dynamism of scientific knowledge production.

The stage of testing a model (through empirical and/or thought experiments), including the critical analysis of the results and the modification or rejection of a model, is probably the best experience students may have in MBT contexts in order to understand issues concerning the methods and instruments used in science. For instance, it is likely that students get involved in practical work at this stage. As in the creation stage, it will have an investigative nature, that is, it will allow students to realise that, although some procedures used in practical work are based on conventions, it is not a question of following a recipe. This may be so even when the MBT activities require students to perform specific empirical tests because students tend to propose other tests (or modifications of the ones being done) in order to confirm their ideas.

Additionally, in science, the conduct of enquires is influenced by a series of socio-cultural factors (like those identified in the 'Whole Science' and 'Family Resemblance', for example, institution, economics/funding, communication, scientific ethos, social values). Many of the features involved in each of them can also be experienced by students, or just permeate their actions. For instance, students may learn about the importance of:

- collaboration within and between groups, as well as 'competition' among groups (something that also contributes to their emotional involvement in the performance of the activities);

- forms of persuasion, since they have to convince their peers about the validity of a given interpretation of data that supports maintenance or changes in their models;
- having cognitive flexibility so that disagreements when data obtained in tests are interpreted in distinct ways could be resolved;
- source of funding, since they will be able to empirically test their ideas only if the school laboratory is well equipped and they have available time to do this (which is not common in many schools);
- norms of handling scientific data, mainly those related to intellectual honesty, openness to free and critical discussions, and to share ideas, data, etc.;
- respect for the environment, which may be translated into simple actions like the correct disposal of chemicals used in experiments.

When models are being tested, students also defend different models (or ideas related to them) during group or whole class discussions. In the context of such discussions, they can also understand some of the aims and values of scientific knowledge, like:

- the importance of knowledge being consistent, simple, and fruitful;
- the need for knowledge to support clear explanations and significant predictions;
- the value of knowledge's empirical adequacy.

Such aspects can be helpful when selecting or accepting one model from several available. As all these aspects are also valuable when identifying the scope and limitations of a model, when convincing others of the validity of the model, and when persuading others to use a given model, the understanding of such aims and values of science can be enhanced when students participate in activities related to the sub-stage of evaluating a model.

Still in the context of the evaluation of a model, its performance may also contribute to improving students' understanding of other features of science:

- the tentativeness of scientific knowledge;
- the possibility of using multiple models (with distinct explicative and predictive powers) in a given context;
- the relevance of discussions supported by evidence for the acceptance of a given model;
- the various forms of communication of scientific knowledge and their adequacy for different purposes;
- the forms of persuasion and the ethical way to use them in science;
- the purpose of using scientific knowledge to understand natural phenomena in the world as experienced;
- the purpose of using scientific knowledge to understand the products of technological activities.

The outcomes in terms of students' learning about science when participating in MBT activities can be related not only to individual characteristics of the modelling stages, but also to some methodological issues. The main ones are:

- The fact that students work in groups. When working in small groups, in general there are no formal hierarchical relations within them. So, students feel able to express their ideas and doubts, to criticise their peers' ideas and be criticised by them, to build on the ideas of others in order to achieve the same purpose. As a result, they may view the social construction and negotiation of knowledge as 'natural' processes.
- The existence of whole class discussions. Once each group has created a given model (very often from intense discussion), students tend to view it as 'their' model. In the following whole class discussions, they have to express their model in the clearest possible way in order to show its adequacy to the purposes previously defined, as well as to convince colleagues whose groups have produced distinct models that their own model is better. On such occasions, students have to use all the argumentative abilities required for arguing, producing alternative theories, and refuting (listed in Table 6.1). These are experiences involving scientific socio-cultural aspects (norms of communicating knowledge, competition, forms of persuasion, peer review, response to criticism, and resolving disagreement) and, as so, may support students learning about them.

Overall, students' participation in MBT activities in which stages are experienced several times and in different ways may help them to reach the appreciation that modelling is a complex, dynamic, and non-linear process of scientific knowledge building; that it involves – and, in fact, depends on – many other scientific practices; that scientific claims have to be validated from a concise and ethical use of evidence in order to be accepted; that socio-cultural dimensions influence the production and use of scientific knowledge; that scientific knowledge may be essential in solving technological, environmental, and social problems, thus having a potential to contribute to the welfare of human beings. These and other more detailed features, as previously discussed, can characterise students' learning about science. Some of them, as discussed in this section, are derived from students experiencing specific aspects of distinct stages in modelling stages. They are influenced by the modelling approach adopted. For instance, as emphasised in Chaps. 2 and 4, one of the characteristics that distinguish the Model of Modelling approach from other modelling approaches is its explicit focus on the stages of expressing and evaluating a model. Therefore, students' understanding of features of science that are directly associated with these stages (as discussed earlier in this section) may not be supported when they participate in MBT activities from other perspectives. They may be more strongly supported when they participate in activities designed and conducted from the Model of Modelling approach (which provide explicit opportunities to students experience such modelling stages).

On the other hand, the achievement of such a comprehensive learning about science cannot be expected simply from students' participation in a modelling-based teaching sequence for any scientific topic. It seems clear, for instance, that modelling activities focused on very abstract and specific scientific topics (for example, intermolecular interactions) are less likely to contribute to students' learning of aspects like the role of scientific knowledge in solving technological, environmental, and social problems. However, it also seems clear that such learning can be easily supported by students' involvement in modelling authentic (or socio-scientific) contexts. At the time when this book is being written, although some teaching sequences involving modelling of authentic contexts from the Model of Modelling perspective have just been designed, no study was conducted on their application. But the evidence from the a study that investigated modelling of authentic contexts from another approach (Prins, Bulte, & Pilot, 2011) supports our belief that similar successful outcomes could be obtained when authentic context were focused from the Model of Modelling approach.

Finally, we are not naively imagining that students can develop a broad understanding about science from a single participation in a modelling-based teaching sequence. Studies conducted in this kind of context (for instance, those reported in Maia & Justi, 2009; Mendonça & Justi, 2011) show that students can develop a sustained knowledge of some aspects, mainly (i) the tentativeness of models; (ii) the possible existence of multiple models; (iii) the importance of posing questions, making observations, collecting data, formulating hypotheses, creating models, comparing alternative models and deciding in favour of the more relevant one for a given purpose or in a given context; (iv) the relevance of the explicative and predictive power of models; (v) the ethical way of collecting and interpreting data, as well as of arguing in favour or against a given model. Despite this, as modelling was an approach totally new for them at that time, some students exhibited a kind of resistance to accept, for instance, the tentativeness of models, or the need to involve themselves in practices for generating knowledge rather than receiving the 'right answer' from the teacher. On the other hand, the outcomes were different in the 7-months study in which students participated in three teaching sequences based on modelling mentioned in Chap. 5 (Maia, 2009). In that study, 30 % of the students were interviewed in different moments during the academic year. As some of the questions focused on the relationships that they viewed between the process they were experiencing and the production of scientific knowledge, their answers, together with the researcher's observations, made it clear that they have developed a broad understanding about science. This seems to show that the opportunities to get involved in many modelling activities, and to become more motivated to do so after the initial modelling-based teaching sequence, were crucial in supporting students' learning about science.

Contributions from the Teachers' Actions

The contributions from teachers' actions may vary depending on the context in which they occur. Therefore, before discussing them, we characterise the distinct contexts in which students' understanding about science can be supported.

Some Relevant Distinctions from the Literature

In the science education literature, there is an extensive discussion about how NOS should be taught, researchers being divided into those that defend the implicit approach and those that defend the explicit approach (Abd-El-Khalick & Lederman, 2000; Rudge, Cassidy, Furford, & Howe, 2014).

The implicit approach is characterised by students acquiring "NOS understanding as a consequence of engaging in learning activities primarily intended for other purposes" (Wong & Hodson, 2010, p. 1436). In this sense, "an understanding of NOS is a learning outcome that can be facilitated through science process skills instructions, science content coursework, and doing science" (Akerson, Abd-El-Khalick & Lederman, 2000, p. 297–298). An essential aspect is that students develop such understanding on their own, without any external help. The literature also emphasises that the implicit approach was adopted by the science projects designed and largely used in the 1960s and 1970s worldwide (like those produced in the USA: the Chemical Education Material Study, CHEM-Study, the Biological Sciences Curriculum Study, BSCS, and the Physical Science Study Curriculum, PSSC) (Abd-El-Khalick et al., 1998; Khishfe & Abd-El-Khalick, 2002).

For many researchers (Abd-El-Khalick et al., 1998; Akerson, Abd-El-Khalick, & Lederman, 2000; Khishfe & Abd-El-Khalick, 2002), the implicit approach has proved to be ineffective. They affirm that their claim is supported by the results of studies conducted at that time, these having indicated that students who learnt the sciences through such projects did not develop an acceptable understanding of NOS. However, an analysis of two to these old studies published in the same issue of the Journal of Research in Science Teaching shows that their results are not homogeneous. Trent (1965) and Crumb (1965) conducted quantitative comparisons of two samples of physics students (one involved in a PSSC course and the other in a traditional textbook-centred course) by using the same written instrument (the 'Test on Understanding Science', TOUS). Trent's results showed no significant difference between the experimental and the control groups "in mean scores of understanding science as measured by the Test on Understanding Science" (p. 228), that is, the results showed no special contribution of the PSSC course to students' understanding of NOS. On the other hand, Crum's results showed that there is a significant difference in understanding of NOS (in favour of students who attended the PSSC course) when the impact is measured over the short period of one semester. However, it is not clear whether data gathered by the use of TOUS can support discussions on students' understanding of NOS. This is so because, according to Lederman, Wade, and Bell (1998), TOUS is 60-item multiple choice test that allows the calculation of scores related to three scales: "understanding about the scientific enterprise, the scientist, and the methods and aims of science" (p. 598). The characteristics related to NOS are only directly accessed by the third subscale. Moreover, Lederman et al. (1998) affirm that the TOUS items "are not related to a student's conception of scientific knowledge" (p. 599), which they assume to be an essential element of students' NOS understanding. On account of this, this group of researchers are among the stronger defenders of the explicit approach for teaching NOS.

The explicit approach advocates that "NOS should be addressed intentionally and explicitly" (Bell, Lederman, & Abd-El-Khalick, 1998, p. 1060), that is, "the understanding of NOS should be considered a cognitive learning outcome" (Khishfe & Abd-El-Khalick, 2002, p. 553), like those related to other components of the curriculum. The explicit approach also includes a "reflective" component, which requires that the subjects were provided with opportunities to analyse

> the activities in which they are engaged from various perspectives (e.g., a NOS framework), to map connections between their activities and ones undertaken by others (e.g. scientists), and to draw generalizations about a domain of knowledge (e.g. epistemology of science) (Khishfe & Abd-El-Khalick, 2002, p. 555).

This reflective component has been shown as particularly relevant when the subjects are pre- or in-service science teachers (Akerson et al., 2000; Akerson & Hanuscin, 2007; Lederman & Abd-El-Khalick, 1998). In general, teachers are initially introduced to aspects of NOS, and then they are provided with opportunities to reflect on them, which results in the improvement of their NOS views. From the results of the studies mentioned in this paragraph, their authors claim that the explicit reflective approach is more effective than the implicit one.

We view this comparison as complicated, because the effectiveness of a teaching approach depends on the aims defined for that teaching, and the measure of its effectiveness depends on the instrument used. In this way, positive results of studies in which (i) tenets from NOS list are explicit taught, and (ii) the subjects' knowledge is accessed through instruments based on the same lists, can certainly indicate that the instruction achieved its aim. However, whether a broad understanding about science is achieved is questionable since, whilst the students may be able to repeat, or even use, the tenets, this does not show a broad view about nature of science.

In relation to the explicit approach, from the analysis of pre-college programmes, Duschl and Grandy (2013) suggest that, in fact, there are two alternative interpretations of the results obtained. What they call Version 1 is based on the explicit discussion of consensus-based heuristic tenets. On the other hand, being based on a comprehensive view of science and science education, according to which cognitive, epistemic, and social practices, as well as technological contexts, conceptualise science,

> Version 2 is grounded in learners experience of 'Building and Refining Model-Based Scientific Practices' in critique and communication enactments that occur in longer immersion units and learning progressions (Duschl & Grandy, 2013, p. 2109).

This means that, when involved in Version 2 approaches, students engage in scientific practices focused on the construction of models for a considerable period of time (weeks or months). In other words, students learn about science through experiencing it as members of a community,

> being immersed in the cognitive, epistemic and social enactments and practise of science that involve building and refining questions, measurements, representations, models, and explanations (Duschl & Grandy, 2013, p. 2126).

In this sense, what seems to differentiate implicit approaches from Version 2 explicit approaches is the kind of activities in which students will be engaged. In implicit approaches, they are of a more general nature, that is, even when some scientific practices are involved, the experience of such practices are not the focus of the teaching as it occurs in Version 2 explicit approaches. Additionally, as science teaching explicitly focused on scientific practices is an emerging research focus in science education, the extent at which Version 2 explicit approaches really contribute to the development of students' views about science is being questioned.

Teachers' Contributions to Students Learning About Science in MBT Contexts

As emphasised in many chapters of this book, MBT from the Model of Modelling approach fosters students' engagement in many scientific practices. In all the teaching sequences developed up to the date at which this book is being written, such engagement occurs for around 10–18 h (which, depending on the number of classes per week of the schools,[1] are spread over 3–9 weeks). Therefore, such MBT contexts could, in principle, be viewed as candidates to support a Version 2 explicit approach for teaching about science. Many aspects about science (mainly those concerning cognitive and epistemic dimensions of science) are clearly experienced (very often many times) by students when participating in the modelling-based activities. But this means neither that all students abstract the elements that are constituents of such dimensions from their experiences, nor that they realise relationships between what they are doing and nature of science. Therefore, the extent at which specific aspects about science may really become explicit to students depends on a series of factors, among them being the teachers' actions.

From a consideration of teachers' actions in MBT contexts based on the Model of Modelling approach (as detailed in Chap. 11), we can identify two levels of teachers' action that may contribute to students learning. One is related to the conduct of the modelling activities and the other related to any special emphasis that the teacher may provide in the discussions that took place in the classes.

One of the main roles a teacher plays when conducting MBT activities is to ensure that students are engaged in all the stages of modelling, which means that they have the opportunity to use the skills and abilities listed in Tables 4.1, 6.1, and 7.1. Therefore, it is essential that teachers explicitly support the use and/or development of such skills and abilities. By so doing, they should, it is hoped, contribute to students' learning of some of the dimensions of science (Tables 9.1 and 9.2), mainly the cognitive and epistemic ones. For instance, teachers may emphasise: the relevance of each of the features concerning process of enquiry at particular moments; the centrality of the purpose for which a model is being produced by guiding students' thinking and actions during the whole process; the importance of using sci-

[1] In Brazil and Mexico, where such teaching units have already been used in regular classes, the number of classes per week for each discipline varies depending on the discipline and the type of the school. Scientific disciplines have, in general, 2–4 (40–50 min) classes a week.

entific and ethical values as well as norms of thinking when designing and conducting data gathering and analysis.

The MBT process is always permeated by discussion. On many occasions, students discuss with their colleagues from the same group. On other occasions, the teacher participates in each group's discussions and coordinates whole-class discussions. On such occasions, beyond emphasising the features mentioned in the previous paragraph, the teacher can also bring to the discussion the features of the socio-cultural dimension – most of which cannot be directly experienced by students when undertaking MBT activities. For instance, in the previous section we affirmed that, when conducting practical work, students might learn about the importance of the funding of science, since this kind of activity can only be performed if the necessary equipment, chemicals, and time were available. However, most students may study in schools without the material conditions that make such activities possible. But the teacher may focus on the funding issue by discussing historical examples of the work of scientists investigating the same topic being learnt by students in which specific funding was essential. When students are conducting experiments in the 'chemical equilibrium modelling' activities discussed earlier, for instance, the teacher may tell and discuss with them some parts of the history of Fritz Haber. Haber's studies around the synthesis of ammonia from hydrogen and nitrogen gases culminated in the proposition of a method that could be performed at industrial scale, and that supported the conversion of ammonia into ammonium sulphate or nitric acid and nitrates (compounds essential to the Germans in the context of the First World War). Part of his studies were possible only from the support of the German company BASF, where he also worked with other scientists (mainly Carl Bosch), who provided significant contributions to the development of his process by making it possible to be performed on an industrial scale. Thus, the inclusion of this historical case in a whole class discussion may help students to understand not only the crucial role of funding in the development of scientific knowledge, but also:

- the relevance of collaboration and competitions among scientists, since for many years many scientists and students collaborated with Haber and/or disputed the authorship of some of the ideas associated with his name;
- the existence of more broad relationships between science and economics. In the case of Haber, his studies were essential to Germany. At that time, Germany had no natural source of nitrates, which were used to produce explosives and soil fertilizers, and could no longer import guano (a natural source of nitrates) from Chile (from where it was originated) because the British Royal Navy were in control of the long sea-lanes from South America. In fact, it seems reasonable to imagine that the First World War would not have lasted so long if Germany had not been so able to manufacture explosives;
- how ethical values may influence the production, acceptance, and use of scientific knowledge. The outcomes of Haber's work could be used to improve agriculture (since soil fertilizers were essential to improve the growing of food) and to the manufacture of explosives. In fact, for some time the declared motivation

for the conduct of the work was the search for a way to reduce nitrogen to ammonia in order to use the product in the production of fertilizers. However, facts like the major use of the products developed in the manufacturing of explosives, and Haber's personal involvement in the War (he abandoned all research not related to the War and personally participated in real tests of chlorine – an irritant gas that causes death from pulmonary oedema – as a weapon) contributed to make his Chemistry Nobel Prize laureateship in 1918 controversial, even among members of the Nobel Foundation (Huxtable, 2002; Wisniak, 2002).

The use of historical cases in order to support the learning of aspects about science has been advocated by many science educators (for instance, Allchin, 2012a; Braga, Guerra, & Reis, 2012; Hodson, 2008; Matthews, 1991, 1994), whose studies have making the success of this approach evident. In particular, the discussion of the Haber case could be very useful in a study aimed at developing pre-service teachers' knowledge about science (Justi & Mendonça, 2014). Evidence from historical case-based studies (like those discussed in the papers mentioned in this paragraph) supports our prediction that, when teachers use them (or parts of them) to base discussions in the MBT context, students may improve their understanding of some dimensions and features of science.

Contributions from the Whole MBT Approach

The example briefly commented on in the last sub-section shows how particular emphases given by teachers in class discussions may contribute to the enhancement of students' understanding about science when conceived within broad perspectives like that of the 'Family Resemblance' and the 'Whole Science' approaches. The existence of such emphases in MBT contexts means that the explicit inclusion of some features of science in class discussions could be viewed as a Version 1 explicit approach to teach about nature of science. However, it is clearly different from the Version 1 explicit approaches as discussed in the literature because each specific feature or dimension of science is contextualised in terms of the MBT activities, that is, they are related to what students are doing or thinking in the modelling process.

On the other hand, MBT can also be viewed as a Version 2 explicit approach to teaching about nature of science, since students are engaged in many scientific practices, and may understand many issues about nature of science from their own experience. Therefore, it seems we are faced with an alternative approach – a Version 3 explicit one – according to which students are simultaneously engaged in scientific practices and given opportunities to participate in explicit discussions of features and dimensions of science related to them and considered relevant by the teacher. Or, in other words, it is a case in which one approach complements the other, as recently reported in the literature (Allchin et al., 2014). As modelling-based teaching sequences are not structurally closed, that is, the nature of the activities and practices provided to students vary depending mainly on the topic studied, the particular science from which such topic is approached, and the goals of the teaching at that particular school level, the simultaneous focus on aspects about science may also support students recognition of diversity among sciences and among scientists way of working (as clearly defended by the 'Family Resemblance' approach).

By characterising MBT from the Model of Modelling approach as a Version 3 explicit approach to teaching about science, we are also assuming 'learning about science' is as characterised by Hodson (2014a, 2014b) or by recent official guidelines for science education (National Research Council, 2012), that is, as a broad perspective that includes cognitive, epistemic, and socio-cultural dimensions of scientific knowledge and practices. Being consistent with this option, we recognise the need to evaluate students' understanding about science from instruments similar to the ones proposed by Allchin (2011, 2012a, 2013), that is, by requesting students to use their acquired knowledge about science in the analysis of contemporary or historical cases. As this is a task we has not carried out yet, we were careful when writing this chapter and opted to use modal verbs when making reference to students learning about science in MBT contexts.

References

Abd-El-Khalick, F. (2012). Examining the sources for our understandings about science: Enduring conflations and critical issues in research on nature of science in science education. *International Journal of Science Education, 34*(3), 353–374.

Abd-El-Khalick, F. (2013). Teaching with and about nature of science, and science teacher knowledge domain. *Science & Education, 22*(9), 2087–2107.

Abd-El-Khalick, F., Bell, R. L., & Lederman, N. G. (1998). The nature of science and instructional practice: Making the unnatural natural. *Science Education, 82*(4), 417–436.

Abd-El-Khalick, F., & Lederman, N. G. (2000). The influence of history of science courses on students' views of nature of science. *Journal of Research in Science Teaching, 37*(10), 1057–1095.

Akerson, V. L., Abd-El-Khalick, F., & Lederman, N. G. (2000). Influence of a reflective explicit activity-based approach on elementary teachers' conceptions of nature of science. *Journal of Research in Science Teaching, 37*(4), 295–317.

Akerson, V. L., & Hanuscin, D. L. (2007). Teaching nature of science through inquiry: Results of a 3-year professional development program. *Journal of Research in Science Teaching, 44*(5), 653–680.

Allchin, D. (2011). Evaluating knowledge of the nature of (Whole) science. *Science Education, 95*(3), 518–542.

Allchin, D. (2012a). The minnesota case study collection: New historical inquiry case studies for nature of science education. *Science & Education, 21*(9), 1263–1281.

Allchin, D. (2012b). Toward clarity on whole science and KNOWS. *Science Education, 96*(4), 693–700.

Allchin, D. (2013). *Teaching the nature of science: Perspectives & resources*. Saint Paul, MN: SHiPS Educational Press.

Allchin, D., Andersen, H. M., & Nielsen, K. H. (2014). Complementary approaches to teaching nature of science: Integrating student inquiry, historical cases, and contemporary cases in classroom practice. *Science Education, 98*(3), 461–486.

Alters, B. J. (1997a). Nature of science: A diversity or uniformity of ideas? *Journal of Research in Science Teaching, 34*(10), 1105–1108.

Alters, B. J. (1997b). Whose nature of science? *Journal of Research in Science Teaching, 34*(1), 39–55.

Bell, R. L., Lederman, N. G., & Abd-El-Khalick, F. (1998). Implicit versus explicit nature of science instruction. An explicit response to Palmquist and Finley. *Journal of Research in Science Teaching, 35*(9), 1057–1061.

Braga, M., Guerra, A., & Reis, J. C. (2012). The role of historical-philosophical controversies in teaching sciences: The debate between Biot and Ampère. *Science & Education, 21*(6), 921–934.

Cakici, Y., & Bayir, E. (2012). Developing children's views of the nature of science through tole play. *International Journal of Science Education, 34*(7), 1075–1091.

Chalmers, A. F. (1982). *What is this thing called Science?* (2nd ed.). Milton Keynes, UK: Open University Press.

Crumb, G. H. (1965). Undertanding of science in high school physics. *Journal of Research in Science Teaching, 3*(3), 246–250.

Deng, F., Chen, D.-T., Tsai, C.-C., & Chai, C. S. (2011). Students' views of the nature of science: A critical review of research. *Science Education, 95*(6), 961–999.

Department for Education. (2014). *The national curriculum in England – framework document.* London, UK: Department for Education.

Donnelly, J. (2001). Contested terrain or unified project? 'The nature of science' in the national curriculum for England and Wales. *International Journal of Science Education, 23*(2), 181–195.

Driver, R., Leach, J., Millar, R., & Scott, P. (1996). *Young people's images of science*. Buckingham, UK/Philadelphia, PA: Open University Press.

Duschl, R., & Grandy, R. E. (2013). Two views about explicitly teaching nature of science. *Science & Education, 22*(9), 2109–2139.

Duschl, R., Schweingruber, H. A., & Shouse, A. W. (2007). *Taking science to school: Learning and teaching science in grades K-8*. Washington, DC: National Academies Press.

Eastwood, J. L., Sadler, T. D., Zeidler, D. L., Lewis, A., Amiri, L., & Applebaum, S. (2012). Contextualizing nature of science instruction in socioscientific issues. *International Journal of Science Education, 34*(15), 2289–2315.

Eflin, J. T., Glennan, S., & Reisch, G. (1999). The nature of science: A perspective from the philosophy of science. *Journal of Research in Science Teaching, 36*(1), 107–116.

Elby, A., & Hammer, D. (2001). On the substance of a sophisticated epistemology. *Science Education, 85*(5), 554–567.

Erduran, S., & Dagher, Z. R. (2014). *Reconceptualizing the nature of science for science education – scientific knowledge, practices and other family categories*. Dordrecht, The Netherlands: Springer.

Grandy, R. E., & Duschl, R. (2007). Reconsidering the character and role of inquiry in school science: Analysis of a conference. *Science & Education, 16*(2), 141–166.

Hodson, D. (2008). *Towards scientific literacy: A teachers' guide to the history, philosophy and sociology of science*. Rotterdam, The Netherlands: Sense.

Hodson, D. (2009). *Teaching and learning about science: Language, theories, methods, history, traditions and values*. Rotterdam, The Netherlands: Sense.

Hodson, D. (2014a). Learning science, learning about science, doing science: Different goals demand different learning methods. *International Journal of Science Education, 36*(15), 2534–2553.

Hodson, D. (2014b). Nature of science in the science curriculum: Origin, development, implications and shifting emphases. In M. Matthews (Ed.), *International handbook of research in history, philosophy and science teaching* (pp. 911–970). Dordrecht, The Netherlands: Springer.

Huxtable, R. J. (2002). Reflections: Fritz Haber and the ambiguity of ethics. *Proceedings of the Western Pharmacology Society, 45*, 1–3.

Irzik, G., & Nola, R. (2011). A family resemblance approach to the nature of science for science education. *Science & Education, 20*(7–8), 591–607.

Irzik, G., & Nola, R. (2014). New directions for nature of science research. In M. Matthews (Ed.), *International handbook of research in history, philosophy and science teaching* (pp. 999–1021). Dordrecht, The Netherlands: Springer.

Justi, R., & Mendonça, P. C. C. (2014). *Contributions of the discussion of a controversy related to history of science to the development of science teachers' knowledge about science*. Paper presented at the III international history, philosophy, and science teaching latin American conference, Santiago, Chile.

Khishfe, R., & Abd-El-Khalick, F. (2002). Influence of explicit and reflective versus implicit inquiry-oriented instruction on sixth graders' views of nature of science. *Journal of Research in Science Teaching, 39*(7), 551–578.

Lederman, N. G. (2006). Syntax of nature of science within inquiry and science instruction. In L. B. Flick & N. G. Lederman (Eds.), *Scientific inquiry and nature of science* (pp. 301–317). Dordrecht, The Netherlands: Springer.

Lederman, N. G., & Abd-El-Khalick, F. (1998). Avoiding de-natured science: Activities that promote understandings of the nature of science. In W. F. McComas (Ed.), *The nature of science in sciecne education: Rationales and strategies* (pp. 83–126). Dordrecht, The Netherlands: Kluwer.

Lederman, N. G., Wade, P. D., & Bell, R. L. (1998). Assessing the nature of science: What is the nature of our assessment? *Science & Education, 7*(6), 595–615.

Maia, P. F. (2009). *Habilidades Investigativas no Ensino Fundamentado em Modelagem [Investigative skills in modelling-based teaching]*. PhD thesis, Universidade Federal de Minas Gerais, Belo Horizonte, Brazil.

Maia, P. F., & Justi, R. (2009). Learning of chemical equilibrium through modelling-based teaching. *International Journal of Science Education, 31*(5), 603–630.

Matthews, M. R. (1991). *History, philosophy, and science teaching: Selected readings*. Toronto, ON/New York, NY: OISE and Teachers College Press.

Matthews, M. R. (1994). *Science teaching: The role of history and philosophy of science*. New York, NY/London, UK: Routledge.

Matthews, M. R. (1998). In defense of modest goals when teaching about the nature of science. *Journal of Research in Science Teaching, 35*(2), 161–174.

Matthews, M. R. (2012). Changing the focus: From nature of science to features of science. In M. S. Khine (Ed.), *Advances in nature of science research* (pp. 3–26). Dordrecht, The Netherlands: Springer.

McComas, W. F. (2008). Seeking historical examples to illustrate key aspects of the nature of science. *Science & Education, 17*(2–3), 249–263.

McComas, W. F., & Olson, J. K. (1998). The nature of science in international science education standards documents. In W. F. McComas (Ed.), *The nature of science in science education* (pp. 41–52). Dordrecht, The Netherlands: Kluwer.

Mendonça, P. C. C., & Justi, R. (2011). Contributions of the Model of Modelling diagram to the learning of ionic bonding: Analysis of a case study. *Research in Science Education, 41*(4), 479–503.

Millar, R., & Osborne, J. (1998). *Beyond 2000: Science education for the future*. London, UK: King's College, London School of Education.

National Research Council. (2012). *A framework for K-12 science education: Practices, crosscutting concepts, and core ideas*. Washington, DC: The National Academies Press.

Osborne, J. (2014). Teaching scientific practices: Meeting the challenge of change. *Journal of Science Teacher Education, 25*(2), 177–196.

Osborne, J., Collins, S., Ratcliffe, M., Millar, R., & Duschl, R. (2003). What "Ideas-about-Science" should be taught in school science? A Delphi study of the expert community. *Journal of Research in Science Teaching, 40*(7), 692–720.

Prins, G. T., Bulte, A. M. W., & Pilot, A. (2011). Evaluation of a design principle for fostering students' epistemological views on models and modelling using authentic practices as contexts for learning in chemistry education. *International Journal of Science Education, 33*(11), 1539–1569.

Rudge, D. W., Cassidy, D. P., Furford, J. M., & Howe, E. M. (2014). Changes observed in views of nature of science during a historically based unit. *Science & Education, 23*(9), 1879–1909.

Rudge, D. W., & Howe, E. M. (2009). An explicit and reflective approach to the use of history to promote understanding of the nature of science. *Science & Education, 18*(5), 561–580.

Ryder, J., Leach, J., & Driver, R. (1999). Undergraduate science students' images of science. *Journal of Research in Science Teaching, 36*(2), 201–219.

Schwartz, R. S., & Crawford, B. A. (2006). Authentic scientific inquiry as context for teaching nature of science. In L. B. Flick & N. G. Lederman (Eds.), *Scientific inquiry and nature of science* (pp. 331–355). Dordrecht, Netherlands: Springer.

Schwartz, R. S., Lederman, N. G., & Abd-El-Khalick, F. (2012). A series of misrepresentations: A response to Allchin's whole approach to assessing nature of science understandings. *Science Education, 96*(4), 687–692.

Schwartz, R. S., Lederman, N. G., & Crawford, B. A. (2004). Developing views of nature of science in an authentic context: An explicit approach to bridging the gap between nature of science and scientific inquiry. *Science Education, 88*(4), 610–645.

Smith, M. U., Lederman, N. G., Bell, R. L., McComas, W. F., & Clough, M. P. (1997). How great is the disagreement about the nature of science: A response to alters. *Journal of Research in Science Teaching, 34*(10), 1101–1103.

Tolvanen, S., Jansson, J., Vesterinen, V.-M., & Aksela, M. (2014). How to use historical approach to teach nature of science in chemistry education. *Science & Education, 23*(8), 1605–1636.

Trent, J. (1965). The attainement of the concept "Understanding Science" using contrasting physics courses. *Journal of Research in Science Teaching, 3*(3), 224–229.

van Dick, E. M. (2011). Portraying real science in science communication. *Science Education, 95*(6), 1086–1100.

Wisniak, J. (2002). Fritz Haber – a conflicting chemist. *Indian Journal of History of Science, 37*(2), 153–173.

Wong, S. L., & Hodson, D. (2009). From the horses' mouth: What scientists say about scientific investigation and scientific knowledge. *Science Education, 93*(1), 109–130.

Wong, S. L., & Hodson, D. (2010). More from the horse's mouth: What scientists say about science as a social practice. *International Journal of Science Education, 32*(11), 1431–1463.

Chapter 10
Learning Progression During Modelling-Based Teaching

Abstract Students will inevitably vary in the rate at which they become experts in modelling, that they acquire meta-modelling capability. If this variation is to be accommodated, the compulsory science curriculum must be structured so as to facilitate 'learning progression' (LP) – the progress to expert status – in some way. The nature of a generic LP is presented that addresses both models and modelling. The attainment of an LP in models and modelling will be intertwined with an LP for each of visualisation, analogy, argumentation, and learning about science. Whilst the detailed structure and testing of such an LP has yet to be done, the issues associated with identifying suitable phenomena to be modelled, with gaining access to such phenomena, and with ensuring that transfer of learning occurs between modelling activities, can be discussed. Finally, the core issue of assessing what progression has taken place at any one time is confronted.

Achieving Worthwhile Learning During MBT

A teacher engaged in MBT would hope that the outcome in the finite amount of time available for such an activity would be that all the students had quickly learned how to produce and test models of phenomena that are of scientific interest. Alas, this ambition is rarely attained, quickly and at expert level, because the acquisition of the knowledge and skills entailed in the development of *meta-capability* is not a unitary process: it actually consists of the development of five distinguishable yet highly inter-related meta-capabilities. An LP in *models and modelling* must be supported by those in *visualisation*, *analogy*, *argumentation*, and in respect of *understanding about science*. The evidence from research is that students show different rates of *progression* in attaining these capabilities, even under the most advantageous circumstances. Moreover, the teacher must seek to enhance all five of them in tandem during MBT if the desired levels of expertise are to be realistically aspired to.

In order to see how this might be done, even in outline, it is necessary to establish what 'progression' entails, before describing what the attainment of each of the five meta-capabilities involve in as much detail as current research has made evident. In respect of the LP in models and modelling itself, two approaches can be outlined. They involve respectively: the provision of a set of models- and modelling-related

J.K. Gilbert, R. Justi, *Modelling-based Teaching in Science Education*, Models and Modeling in Science Education 9, DOI 10.1007/978-3-319-29039-3_10

activities, culminating in the construction of a model *de novo*; and the adoption of a model-based approach to curriculum design.

The Notion of Progression in Learning

The intention behind the provision of an LP in each of the major concepts and skills-sets of science at school level is to support the acquisition of 'scientific literacy for all' (see Chap. 1). A major issue in doing so is the need to bring **about** a much greater coherence in general than at present between what is taught about models and modelling (the requirements of the curriculum), how it is taught (the pedagogy adopted to do so) and the way that the learning that takes place is assessed (the testing regime adopted) (Duschl, Maeng, & Sezen, 2011). Achieving such a synergy from the present very partial and fragmented provision can be seen as one of the major challenges to be faced by science education over the next decade (and probably more!). The development of such an LP is of great importance because, like LPs on other aspects of science (for example, how scientific methodology has developed historically), it will

> … have the potential to organise standards, assessments, and instruction in a way that promotes scientific literacy. Current (…) curricula prioritise the structure of the scientific disciplines, using a top-down approach that creates logical (from scientists' perspective) sequence of ideas. Learning progressions, which use both top-down and bottom-up 'design' approaches, can combine ideas about scientific disciplines with understanding of how students learn (Gotwals & Alonzo, 2012, p. 4).

Thus, while the LP will assume the validity of the current view of models and modelling that is deemed desirable for school-level education, the fact that students have their own interpretations of such matters (as shown, for example, in Grosslight, Unger, Jay, & Smith, 1991) will be recognised at least in the pedagogy adopted.

In somewhat greater precision, the report of a conference by Corcoran and Silander (2009), summarised by Duschl et al. (2011), posited that LPs in general will have four features that permit age- and attainment-related learning in science education to be facilitated by:

> 1. targeting core and generative disciplinary understandings and practices that merge science content with science practices;
> 2. (establishing) lower and upper boundaries that describe entry assumptions and exiting expectations for knowing and doing;
> 3. descriptions of LPs that inform progress levels or steps of achievement;
> 4. (pointing to) purposeful curriculum and instruction that mediates targeted student outcomes. (Duschl et al., 2011, p. 136)

As we have discussed in Chaps. 2 and 9, modelling is a complex operation that is a vital component in the conduct of scientific enquiry. If an authentic science education is to consist of a series of LPs concerned with specific and scientifically important sets of knowledge and skills, leading to competences in them, then we would argue that an LP on modelling will be a vital component of such an education for all students. Such an LP would therefore have, as its core aim, the attainment of competence in modelling. As has already been said, the development (or exercise) of competences in visualisation, analogy, and the nature of science would necessarily also be required in order for this to happen.

Progression in Models and Modelling

The Nature of a Competence in Models and Modelling

Progression in the knowledge and skills required for modelling necessarily entails progression in knowledge about the nature of models. Thus an overall competence in models and modelling will consist of three elements. These are:

- *Knowledge about models.* This is comprised of scientifically acceptable knowledge of: the epistemological and ontological nature of models; the reasons why they are constructed and used; how their scientific value can be assessed.
- *Knowledge about modelling.* This is comprised of scientifically acceptable knowledge of: the epistemological and ontological grounds on the basis of which models can be constructed; the procedures involved in constructing models; the procedures involved in evaluating the procedures involved in the construction of models.
- *Skill in the practice of modelling.* The emphasis here is on what is actually done when a person is actively engaged in the practice of modelling. These elements are manifest in the skills deployed when:

  - Students *construct* models consistent with prior experience and theories to illustrate, explain, or predict phenomena;
  - Students *use* models to illustrate, explain, or predict phenomena;
  - Students compare and *evaluate* the ability of different models to accurately represent and account for patterns in phenomena and to predict new phenomena;
  - Students *revise* models to increase their explanatory and predictive power, taking into account additional evidence or aspects of a phenomenon (Schwarz et al., 2009, p. 635, original emphases)

Evidence on the Attainment of Competence in Modelling

Until the last 20 years or so, the theme of models and modelling was not discussed to any great extent in science education. This neglect is reflected in the wide variety of treatments of the themes themselves in textbooks for students over the years, for example in physics (Niss, 2009).

An early and influential enquiry by Grosslight et al. (1991) concerned with the understanding of the nature of models held by students across a range of ages between those of high school and of university suggested that they attained one of three 'levels' of understanding. At Level 1, a model was seen as a simple copy of some aspect of reality, while at Level 2 it is basically a copy of reality but produced for a specific purpose, and at Level 3 a model is a construction produced to develop and test ideas, not being a copy of reality. The majority of the school-age students in that study displayed an understanding at the first of these three levels. Later work in the same vein by Chittleborough, Treagust, Mamiala, and Mocerino (2005) showed that the understanding of similarly-aged students had shifted somewhat towards the third, the more scientifically-acceptable, level of understanding. This shift may well have been associated with the growing research, development and teaching interest in the field over the period 1991–2005.

Gobert and Pallant (2004) noted that the work that has since taken place has fallen into three phases in each of which enquiry into understanding has been conflated with attempts to develop that understanding. In the first phase, students were presented with a model and required to learn factual content with its use. In the second phase, students were presented with a problem and required to construct models with which to explain it. In the third phase, an extension of the second, the processes of model construction and use have been defended in discussion with students' peers.

The more recent work has emphasised the value of modelling *per se*, this subsuming the development of an understanding of models per se (Schwarz et al., 2009). Situations where the emphasis is on students creating their own models have been found to be especially successful (Abell & Roth, 1995; Lehrer & Schauble, 2012). Closer examination of what is happening during this process of creation, in the context of the use of computer-based modelling systems, showed that students tended to pay attention to the empirical data on the phenomenon in question that is available rather than to their relevant prior knowledge (Sins, Savalsberg, & van Joolingen, 2005). This suggested that there is a clear and major role for teachers in supporting the modelling process if student-initiated work was to be successful. Other work has shown the importance of the adoption of pedagogic practices that closely reflect those in use in scientific enquiry itself (Prins, Bulte, & Pilot, 2011).

The work summarised above does point to the elements of a potentially successful LP. It must: address both models and modelling; be based on an address to phenomena that students see (or merely passively accept) as being of authentic interest; focus on the construction and evaluation of models by the students themselves; involve the provision of scaffolding for student activity by the teacher; involve students reflecting on what they have done during the process of modelling.

Progression in Visualisation

The Nature of Competence in Visualisation

In Chap. 7, we explored the fact that access to the world-as-experienced is provided by the five senses: sight, hearing, smell, taste, touch. From these five, discrete focuses of perception can be identified which have a particular salience. These involve the use of: gesture, concrete objects, vision (especially pictures, diagrams, graphs), and speech. Each of them, together with the many sub-modes in which they exist, differing only in the emphasis placed on their defining attributes, provides particular visualisations. That is, they provide particular links between mental models and the aspect of reality that they depict. The conventions governing the nature of these links are called the 'codes of representation' for the modes or sub-modes.

A science student with competence in visualisation fully understands and can use the modes and sub-modes of representation commonly used in science. To do this, an individual must be able to demonstrate:

- a full understanding of the codes of representation of all the modes and sub-modes of visualisation used in science education. This knowledge will include that about: the entities within models that can be depicted; the relationships between them that can be shown (for example, relative distance and angle between them); the relationships between them that cannot be shown;
- a capacity to 'translate' a given model between those modes of visualisation that are appropriate. For example, that of the human heart between a diagram of it and a material model of it;
- a capacity to construct a visualisation in any one of the modes and sub-modes used in science education;
- a capacity to use a suitable visualisation to solve novel problems in respect of a given phenomenon (Gilbert & Eilam, 2014).

Evidence on the Attainment of Competence in Visualisation

Many thousands of studies of competence in visualisation have been reported in the literature. However, they vary widely in terms of: the definition of 'visualisation' used, the particular mode (or combination of modes) of representation focused on, the particular element of knowledge or skill focused on, the methodology used to obtain data (ranging within the spectrum of the qualitative to the quantitative), the age gender and educational background of the students, the environment within which data is collected. Review articles have attempted to provide overviews of parts of this complex field (for example, Hegarty & Waller, 2005; Newcombe &

Learmonth, 2005), with those concerning sex differences being the most readily comprehensible, for example that of Halbern and Collaer (2005). The consequence is that it is not possible to even approximately answer the question 'To what extent and in what ways is the ability to visualise attained by individuals?' The closest approach to doing so was made by Eilam (2012) in identifying what teachers' might do to promote the status of visualisations in respect of their accompaniment to text. In essence, she suggests that students learn most from visualisations in such a context if:

- they actually have to memorise the representations, which can only be done if they have learnt the codes of representation involved;
- they explain a piece of accompanying text by using what is represented in an adjacent visualisation;
- they explicitly explain elements in a visualisation in terms of the meaning attached to that which is represented;
- they seek understanding from a visualisation that extends the meaning derived from accompanying text;
- efforts are being made to extend their general knowledge of that which is being represented.

As emphasised in Chap. 5, studies conducted in MBT (Gilbert, Justi, & Queiroz, 2010; Justi, Gilbert, & Ferreira, 2009) have also shown that students' involvement in modelling-based activities can support the development of their competence in visualisation when the stage of expressing a proto-model in any of the modes of representation is explicitly focused on and extensively discussed.

Progression in Analogical Reasoning

The Nature of Competence in Analogical Reasoning

In Chap. 8, we established the importance of analogy in MBT. Drawing on that analysis, we suggest that competence in analogical reasoning involves the knowledge and ability to:

- gain *access* to memory of a range of domains that have the potential to provide the basis for suitable analogies;
- systematically *map* parts of the base domain onto the corresponding parts of a chosen target domain;
- *evaluate* the match between the two using the criteria of structural consistency and structural focus;
- deduce *inferences* about the target domain, this producing new or enhanced understanding;

- *generalise* those inferences to targets that seem similar to the target domain being immediately addressed;

in all the target contexts likely to be met in science education.

These processes of analogical reasoning are used in all the elements of our model for MBT (see Chap. 2). Thus, during the:

- *Creation* of a model. Having decided on the aims that a model should address, the modeller collects relevant empirical data, identifies a suitable source for the model, and draws the analogy that leads to its creation as a proto-model. This is the central role for analogy in the modelling process;
- *Expression* of a model that has been created. The modeller has to both identify a suitable mode of representation for it and carry out the process of expressing it. In order to do so, an analogy can be drawn to similar contexts where a particular mode of visualisation has been used;
- *Testing* phase. The expressed model has to be included in the design and conduct a thought experiment, these processes being preferably based on an analogy to a suitable pre-existing thought experiment. Where this is successful, the next step may be the conduct of an empirical experiment: analogy is again involved;
- *Evaluation* of the model. This involves the use of the model to represent a phenomenon in a context different from the one for which it was initially produced.

Evidence on the Attainment of Competence in Analogical Reasoning

The drawing of analogies, as a general learning strategy, has a well-recognised role in First Language education. It is in that context that the problems faced by both young children and students have been identified. For example, Goswami (1992) reports that failure to understand analogy has three fundamental roots: a lack of detailed knowledge of a chosen source, an inability to identify those aspects of the source that might appropriately be mapped onto the target; a lack of understanding of how to draw a relationship between relevant aspects of the two domains.

These problems have also been identified in science education (Aubusson, Harrison, & Ritchie, 2006). The research conducted so far has been largely concerned with students' ability to understand the 'teaching analogies' created by their teachers in the context of direct instruction. Whilst the general problem of lack of background knowledge was impossible to directly address, the 'Focus, Action, Reflection (FAR) Guide' for teachers (Treagust, Harrison, & Venville, 1998) has proved an effective way of ensuring that the second and third roots of misunderstanding could be addressed (Harrison & Treagust, 2006).

The little work that has been undertaken into the value of students' own construction of 'teaching analogies' (which we might term 'self-teaching analogies'), those in teacher-driven contexts have been shown to be successful (Aubusson, Treagust, & Harrison, 2009). As discussed in Chap. 8, the much greater emphasis on students' construction of analogies during MBT suggests that the process of development of their ability to do so warrants close attention in itself. The Teaching with Analogy (TWA) model and the FAR Guide should prove valuable ways of systematically introducing students to the ideas surrounding the construction and use of analogies. In fact, when the TWA model was used as a base for the initial activity in Mozzer and Justi's (2012) study, it was found to be very helpful in supporting students' understanding of the meaning of analogy.

Progression in Argumentation

The Nature of Competence in Argumentation

In Chap. 6, we outlined the skills required for individual participation in argumentation. To do so, a person needed to be able: to deal with evidence, to argue, to counter-argue, to produce alternative theories, to refute arguments (see Table 6.1). Being able to do so in a wide variety of situations constitutes competence in argumentation.

Evidence on the Attainment of Competence in Argumentation

Although a great deal of research into argumentation in science education has taken place in the last decade, it is only fairly recently that overviews of the field have begun to appear (Erduran & Jiménez-Aleixandre, 2008; Mirza & Perret-Clermont, 2009). A review of the literature on student learning (Garcia-Mila & Andersen, 2008) reveals that, in general, students argumentative skills are poor until they are specifically addressed. Most importantly, students:

- do not engage with the arguments put forward by their peers during collaborative working. They only consider their own claims, ignoring those of other people with whom they are (in theory at least) working;
- put forward claims without them being accompanied by potentially supportive evidence;
- put forward evidence, when they do so, that which supports their own claims.

Students' normal lack of competence in respect of argumentation suggests that engagement in modelling activities, of the type that we have outlined in MBT, will lead to the gradual development of these skills. Evidence that support this affirmative is discussed in Mendonça and Justi (2013) and in Chap. 8.

Progression in Understanding About Science

The Nature of Competence of Understanding About Science

Chapter 9 outlined the different perceptions of nature of science that might inform school science education. A combination of the 'Family Resemblance' and the 'Whole Science' views does present a compelling account of the field, but implementing this with the school curriculum would require major changes both to syllabuses and to teacher education.

Evidence on the Attainment of Understanding About Science

Given the complexity of the ideas involved, it is not surprising that such research as exists shows that, in the absence of teaching focused on 'understanding about science', only a relatively small proportion of students show an acceptable grasp of those ideas at any one place in the school curriculum (Kang, Scharmann, & Noh, 2005). In a large-scale, interview-based, study Driver, Leach, Millar, and Scott (1996) showed that students aged from 9 to 16 years in the UK demonstrated a slow and uneven progression in successively using three forms of understanding:

- Phenomenon-based reasoning.
  No distinction is made between observation and explanation. The former involves enquiry by carefully observing objects and events, whilst the latter involves a re-description of those objects or events.
- Relation-based reasoning.
  Whilst students do distinguish between observation and explanation, the latter is thought to emerge from generalisations about the former.
- Model-based reasoning.
  Explanations are based on a model of a phenomenon which has a distinct ontological status. A model arises by the act of imagination, whilst the observations made play a substantial role in its acceptance via the making and testing of predictions.

In the light of these results, it would be expected that explicit attempts to teach 'understanding about science' would yield some positive results. Leach, Hind, and Ryder (2003) showed that, for some 16–18 year old senior high school students at least, the insertion of single lessons spread throughout the school year did lead to a distinctive improvement in an understanding of the epistemology of science. In an associated study, Ryder, Leach, and Driver (1999) showed that undergraduate science students who were conducting project work, but without explicit instruction on epistemological matters, showed a significant improvement in their grasp of 'understanding about science'. In the light of these results, it is not surprising that Khishfe

(2008) showed that adopting an inquiry-based approach to teaching, which entailed an explicit address to epistemological issues, led to a improvement in 'understanding about science' on the part of many of the students.

In all these studies it is puzzling why only some students showed the desired change in understanding. A clue may lie in the finding by Hogan (2000) that there seems to often be a distinction between what she terms students' *distal knowledge* of what is entailed in 'understanding about science' – their declarative knowledge of the subject – and their *proximal knowledge* of what is entailed in it – their personal beliefs and commitments. Whilst a failure to grasp the ideas they are taught about 'understanding about science' may be widespread among students – those ideas are multiple, complex, and abstract – it may also be the case that they do not believe what they are being taught. Supporting progression may thus be even more demanding that at first sight seems to be the case. At the same time, there does seem to be confusion about the theoretical foundations about both nature of 'understanding about science' and how to effectively and efficiently evaluate the progression of learning in and about it (Deng, Chen, Tsai, & Chai, 2011).

As has been shown in the previous five sections of this chapter, our grasp of the manner and extent of students' knowledge of models and modelling, visualisation, analogical reasoning, argumentation, and 'understanding about science', is far from complete. However, this must not excuse us from trying to design an LP in modelling based on current 'best practice'.

A Potential Strategy for Designing an LP About Models and Modelling

The major question that has to be addressed here is: What notion of learning should form the basis of an LP? One approach would see all learning as completely idiosyncratic, such that no firm, generalised, structure for an LP can be proposed: this would be what Ford (2015) describes as "hopelessly tailored to fleeting fluctuations of situation and setting" (p. 407). Most importantly, it would leave teachers without any real guidance on how to design lessons. The other approach would be to provide what Ford (2015) calls "context-free accounts (…) that have questionable validity" (p. 407). This approach, which has been found to be generally ineffective in facilitating learning, based on the reflection of a sequence of ideas and activities derived from a simplified view of the history of the development of the notion of models and modelling, has been widely used up to the present. We feel that an approach that lies between these two is likely to be both the most supportive of the attainment of scientific literacy in respect of models and modelling by all students and to be within with 'subject knowledge' and 'pedagogic content knowledge' (Shulman, 1987) of many teachers (but see Chap. 11).

Such an approach would recognise the validity of what has been learnt about learning in science education over the last few decades and which has been assimilated by teachers. Thus, preconceptions/misconceptions by students have been identified for a large number of the concepts used by science (Gilbert & Watts, 1983). It may thus be assumed that students will have some ideas about many of the ideas that comprise an LP on modelling before they are taught them. We can be more certain about what students should know and be able to do when they are fluent in the knowledge and practice of modelling. However, as they will have started to be taught about modelling at very different points in their educational development, we can be far less certain about the status of their relevant knowledge and skills when they start to experience that process. The essence of any curriculum that addresses 'models and modelling' must therefore involve an address to all the elements of competence in modelling at an elementary level, followed by a progressive re-address to those elements at a higher level, perhaps in successive school years. We have identified two possible strategies based on existing best practice that meet these criteria. It first is self-evident: the explicit provision of a progressive exposure to all the components of a capability in models and modelling. The second is wider in ambition: the design and provision of a curriculum based on ideas about models and modelling. We now address these two alternatives.

Model 1: An Explicit and Progressive Exposure to Competence in Modelling

This approach consists of a direct and progressive acquisition of the elements of knowledge and practical skill that comprise a competence in modelling. It is produced by linking together the conduct and outcomes of activities that have, individually or in conjunction with each other and over a period in time, already been reported in the research literature.

As a result of working with teachers to produce a description of a 'modelling environment' that was suitable for (USA) grades K to 6 in respect of precursors to the study of notions of 'evolution', Lehrer and Schauble (2012) identified some principles that seem Competence in Modellingrelevant to all content-themes in the science curriculum (that is, including that of 'models and modelling') and all student age-ranges. Thus, in order to provide a structured knowledge of modelling, investigations should take place:

- using phenomena (systems, contexts, problems) that are capable of supporting enquiries at a progressively greater depth;
- using varied exemplars of such phenomena so that 'transfer of learning' is facilitated;
- in such a way that meaningful questions about the phenomena can readily be raised by both the teacher and the students;
- in such a way that support the development of a repertoire of types of representations (graphs, diagrams etc.);

- in such a way that support the development of progressively improved practice in the use of those types of representation;
- in such a way that support the development of the practice of measurement and hence of the skills of assessment.

Turning these principles into an actual LP in models and modelling could be produced by successively providing students with opportunities to:

Learn and Use a Model of Models and Modelling

The theme of 'models and modelling' is often first met at the beginning of different phases of education. In these circumstances a conservative, transmission, model of teaching must be (alas) assumed for many such occurrences, perhaps augmented by student-led practical work, in either physics, or chemistry or biology. Many of the recorded instances relate to study at university level, although there seems no *a priori* reason why they should not found at school level. Thus, for example:

- In biology, Passmore and Stewart (2002) directly taught several models of evolution, including Darwin's Theory of Natural Selection, to high school students and had them successfully use it to explain a number of phenomena using self-made representations.
- In physics, Redfors and Ryder (2001) taught university students about the structure of metals and found that their understanding was best when the examples chosen were familiar to the students. Arnold and Millar (1996) introduced the notions of heat, temperature, and thermal equilibrium to 12 year-olds and then had apply them in a range of contexts, thus showing the power of a concept to provide generalise understanding.
- In chemistry, Luxford and Bretz (2013) taught university students in the USA about covalent and ionic bonding and subsequently interviewed them where they were found to be better able to show their understanding when using self-made play-dough representations.

In general, this initial phase consists of: direct teaching of a curriculum model by the teacher, augmented by practical work in which students attempt to relate how that model explains the behaviour of phenomenon by use of representations of it (usually in a concrete/material form).

Learning to Revise Models

In this addition to the 'learn and use a model' approach, students are required to learn and subsequently alter a model that explains a simple phenomenon so that it successfully represents more complex phenomena (Halloun, 1998). Here, a model is adapted to address changing purposes, for example having represented the behaviour of a phenomenon then going on to represent the causes of that behaviour.

In this phase, the emphasis is on changing a model so that it can meet revised/altered different purposes.

Learning to Reconstruct a Model

In this approach, students are provided with all the elements of an established model and with a series of questions, the answers to which will lead them to build that model.

Two examples stand out in the literature. Barab, Hay, Barnett, and Keating (2000) provided undergraduate students with the elements of the standard solar system model and had them assemble the components in order to explain various phenomena shown by the Earth-Sun system. The emphasis here is on the construction of a model, probably heavily 'scaffolded' by the teacher. Louca, Zacharia, and Constantinou (2011) familiarised 11–12 year old students with a computer-based modelling system and required them to work in groups. They attempted to recreate models to explain 'the free-fall of an object', 'relative motion' and 'diffusion'. With the teacher providing supportive questions only, activity by the students passed through three phases: the description of the phenomenon (objects were identified); causal reasons for behaviour were speculated on; the construction of 'stories' about what was happening to individual objects in the production of a comprehensive causal explanation.

At this stage of learning about modelling, the main precepts become present. Whilst the teacher may set the problem, the questions to be addressed are identified by the students, as is the proposal of a model, together with the collection, analysis and evaluation of data. Whilst the main elements of modelling are present then, they are carried out within the psychological safety of a known-model of a phenomenon.

Learning to Construct a Model de Novo

This can be based on the 'Model of Modelling' approach that has be presented and discussed in Chaps. 3 and 4. All the criteria for models and modelling are met. As discussed in Chap. 4, this means that students would need to experience all the stages of modelling, starting from doing so in situation involving simpler entities to be modelled and questions related to them to be answered, and gradually moving to situations involving more complex entities and questions. However, by taking into account what is discussed in this chapter, the progression would not only be related to the target of modelling. Perhaps more important than this, the progression concerning learning to construct a model de novo would require a series of simultaneous progressions in the major epistemic practices that permeate modelling (visualisation, analogical reasoning, and argumentation), thus supporting the main cognitive processes involved in performing it (as discussed in Chap. 2). Therefore, students learning would not be focused only on learning a given scientific or

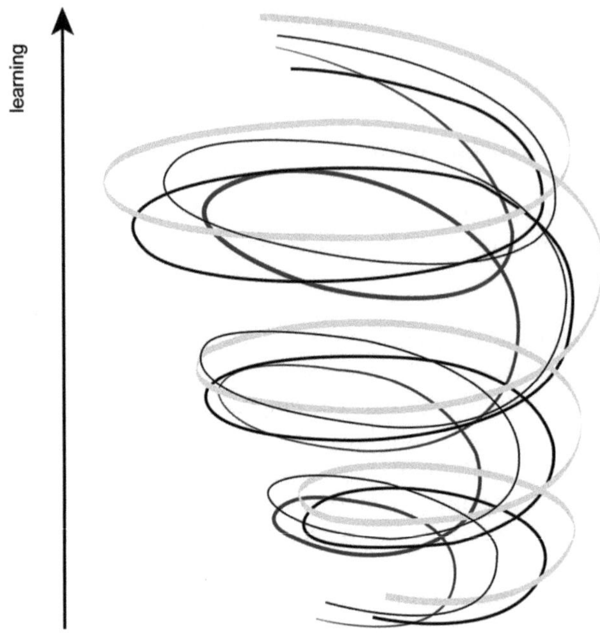

Fig. 10.1 Representation of the progression concerning learning to construct a model de novo

socio-scientific topic through modelling-based activities, but also on simultaneously learning to perform the related epistemic practices. Such progressions would result in (and, in some sense, constitute) progressions in the understanding of the entity being modelled, the development of the competence on models and modelling, and the understanding about science.

In a simplified way, the progression concerning learning to construct a model de novo could be represented through a series of interconnected multiple spirals, each one representing the progression in visualisation, analogical reasoning, argumentation, understanding about science, etc. (Fig. 10.1). It seems impossible to predict or to identify the meanings of each point where two or more spirals cross each other in a theoretical manner or outside a given context. On the other hand, Fig. 10.1 emphasises that:

- the learning process represented is dynamic and non-linear (as it is modelling itself);
- specific points in one spiral may influence and/or support turnings points in a distinct spiral.

The latter aspect was partially observed in some of the studies conducted from the Model of Modelling perspective discussed in Chaps. 4, 5, 6, 7, and 8 when the authors showed, for instance, how the development of students' argumentation supported their better performances during production, expression, tests and evaluation of models (Mendonça & Justi, 2013), or how the development of students' visualisation skills was essential for their learning of both abstract chemical topics and meta-modelling knowledge (Gilbert et al., 2010; Justi et al., 2009; Maia & Justi, 2009). However, in those studies, the empirical data were analysed in order to address questions that did not focus on the relations between distinct L.P. Therefore, this new approach requires other studies from which (i) Fig. 10.1 could be tested, and (ii) the meaning of specific representational aspects in Fig. 10.1 could be better characterised. Certainly, there is a lot yet to be done!

Model 2: Basing the Curriculum Substantially on the Ideas of Models and Modelling

A Basis in Teacher-Led Development Work

This approach has been developed during the work of the 'Cams Hill Science Consortium' which began in 2001 and which continues to evolve to this day. Although only a more detailed account of the first 5 years of the project are available (Newberry & Gilbert, 2007), the general outlines of the project have remained the same throughout the whole period. Thus:

- the work covers the whole age range of compulsory science education in England and Wales (5–16 years);
- the basic objective was to support teachers in the professional provision of a science curriculum based on constructivist principles. It must also meet the requirements of the mandatory National Curriculum (which was undergoing a process of change throughout the period 2001–2015);
- participation is by the invitation of teachers who had shown a capability for curricular innovation. Although participation changes somewhat, as teachers changed their schools, 27 schools were directly involved in 2013. Funding was largely provided by participating schools;
- ideas were initiated, developed, and discussed during meetings spread throughout the 'school year';
- between these meeting, the teachers tried out their ideas in the classroom as they saw fit, reporting the outcomes (both positive and negative) to the next meeting of one of the three geographical sectors of the Consortium.

The Outputs of the Consortium

The initial impetus for the Consortium was the belief by the initiators (Matthew Newberry and John Gilbert) that the ideas of models and modelling had much to contribute to a science curriculum that would engage students. The outputs of the Consortium can be summarised under four headings reflecting themes that were addressed roughly sequentially during the project:

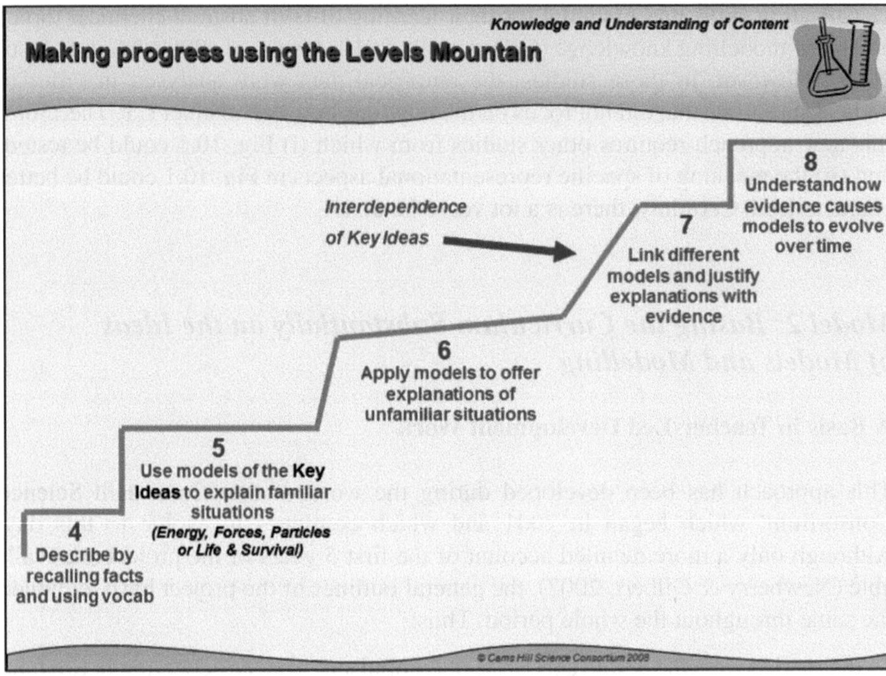

Fig. 10.2 The 'Levels Mountain' representing the increased 'Levels of Understanding' required by the National Curriculum (Newberry, Grevatt, & Gilbert, 2009, p. 21)

1. Providing a coherent basis for the development of understanding.

A careful reading of the National Curriculum convinced the members of the project that the progressive 'levels of attainment' that were expected to be achieved by students were based to Bloom's taxonomy of educational objectives (Bloom, 1956). Being initially focused on the higher 'levels of attainment' (i.e. 4–7), the project initially developed a 'levels mountain', which provided a simple empirical guide simultaneously to the nature of the 'steps in understanding' that were expected, the magnitude of the increased cognitive demands that they successively required, and the relative time that the full attainment of each required (see Fig. 10.2).

The Levels Mountain was readily understood by both teachers and students, being widely used to suggest, in classrooms in the south of the UK, what understanding was required if the 'next Level' was to be attained. While intended initially as a guide for teachers in lesson planning, students also found it helpful in grasping what was expected of them.

However, towards the end of the 2000s, the designers of the National Curriculum abandoned the explicit expectation of 'Levels of Understanding' on the grounds that students' parents found the idea difficult to understand. In order to accommodate the change in terminology that ensued, whilst retaining the notion of models in an

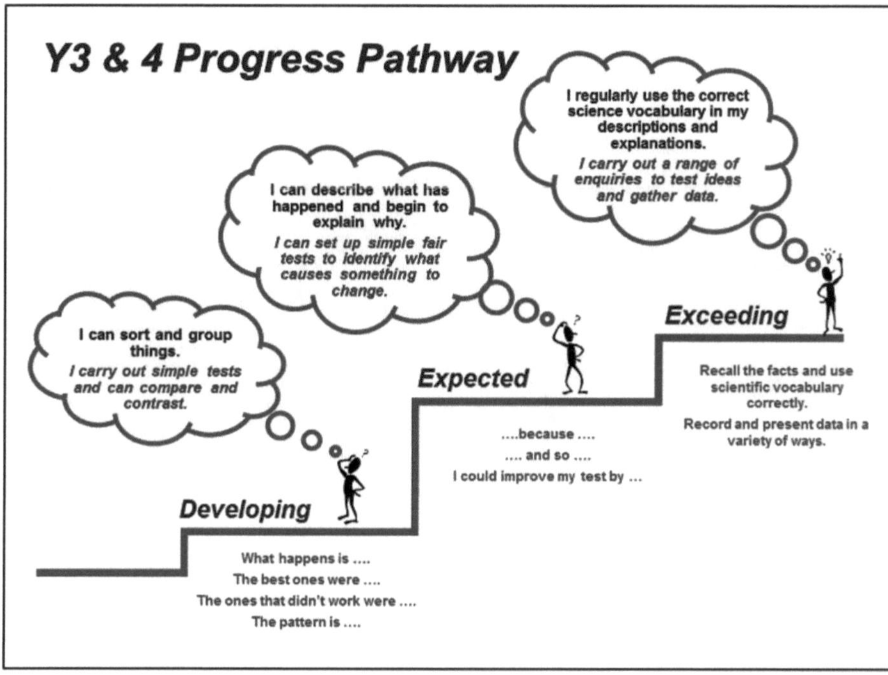

Fig. 10.3 An example (Year 3) of the 'Progress Pathway' for the progression of learning in the Primary School (Newberry & Cams Hill Science Consortium, 2014)

implicit form, the 'Progress Pathway', in which the nature and quality of the explanations that students were expected to understand were set out for each academic year of the primary school (1–5) and across the sum of the first three academic years of the secondary (junior high) school (7–9) (see Figs. 10.3, 10.4.a, and 10.4.b).

Notice that the emphasis has shifted to 'the attainment of explanatory' competence, recognising that, with advancing school years, the pace of learning spreads out across a student cohort whilst the nature of that understanding expected of all is increasingly precise. This approach assumes the notion of 'model' as the driving force in the production of explanations: the more advanced the understanding of the nature of models, the more sophisticated the explanations that are possible.

2. Providing a basis for a constructivist approach to learning about and with models.

The project developed an outline 'scheme of work', based on constructivist principles, that would enable any one of the ways of using models (learning, using, revising, recreating, producing) to be addressed. Called the 'Thinking Frame', its basic structure is given in Fig. 10.5.

An interactive website with further details and videos of the Cams Hill Science Consortium's Thinking Frames Approach can be found on http://pstt.org.uk/ext/cpd/the-thinking-frames-approach.

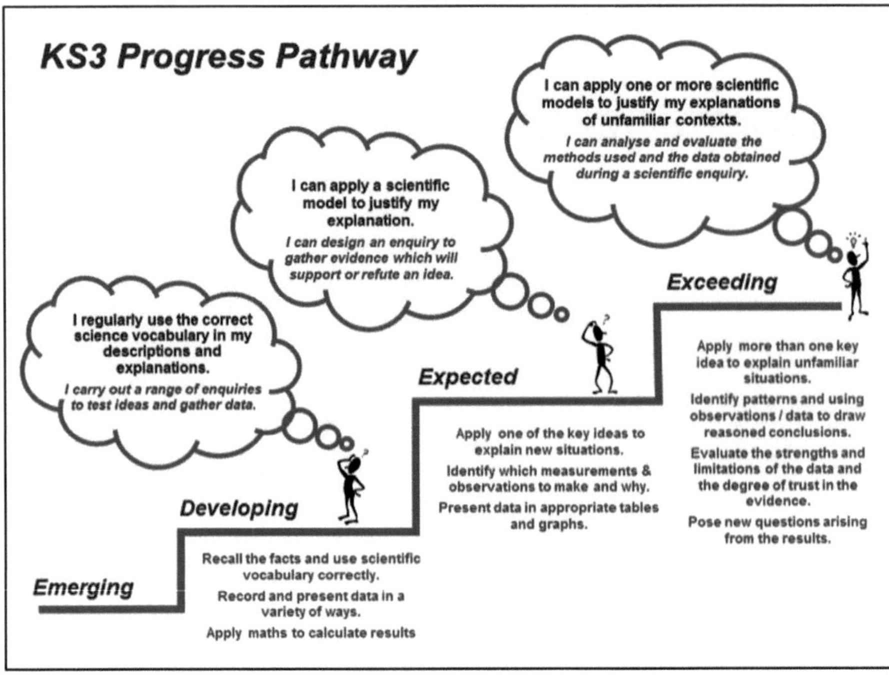

Fig. 10.4.a The 'Progress Pathway' expected to students in Years 7 to 9 (Newberry & Cams Hill Science Consortium, 2014)

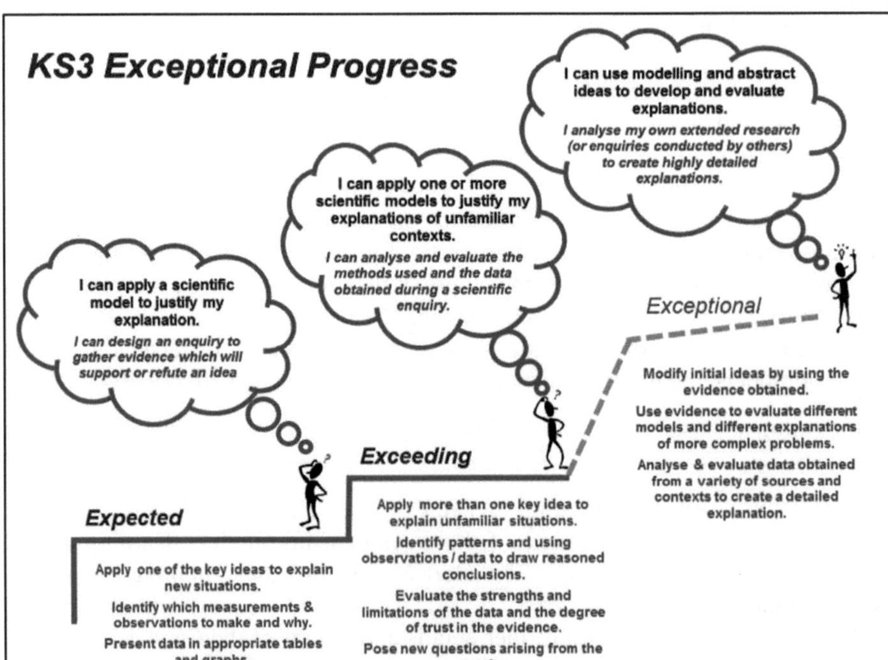

Fig. 10.4.b The 'Exceptional Progress Pathway' designed to challenge able students in Years 8 + 9 (Newberry & Cams Hill Science Consortium, 2014)

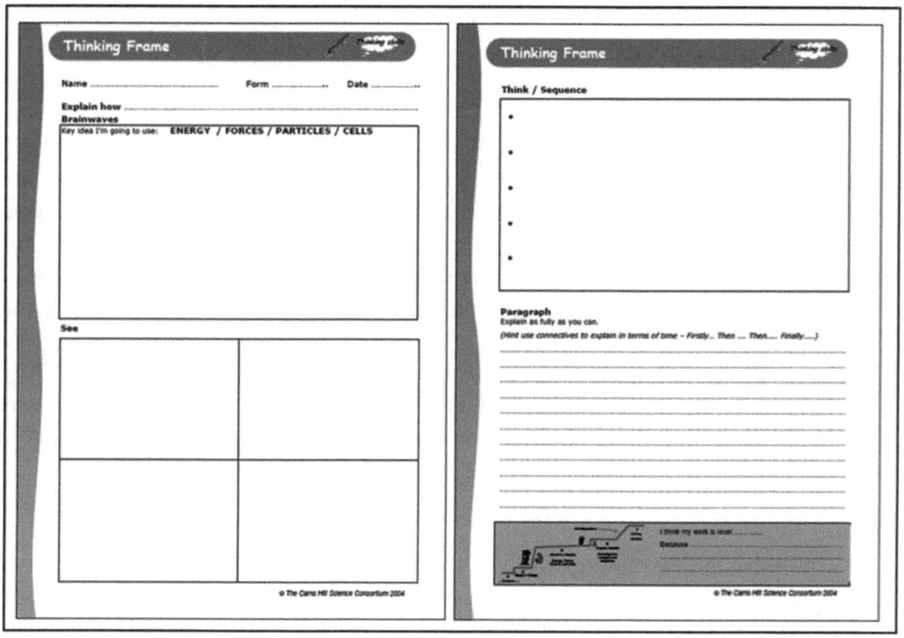

Fig. 10.5 A generic 'Thinking Frame' (Newberry, 2006)

The initial impetus for a class based on the use of a Thinking Frame was the identification of a *problem* associated with a phenomenon that was readily accessible to students. Every effort was to pick on examples of phenomena that were drawn from everyday life: books such as Press (1995) provided the basis for problems as did the 'scientific toys' commercially available, for example 'the sonic gun'.[1] The first step with a class would be elicit *brainwaves* about what was happening in or causing the phenomenon observed. These would be models, whether preconceptions or previously taught models. The next step would be for students to produce a *visualisation* of what was happening or causing the phenomenon. This would involve the various modes of representation. As the use of the Thinking Frame became more common in a given class, students would develop competence in a repertoire of modes. The third step was to produce a *thinking sequence* of a possible explanation as a series of bullet points which was then tested empirically. The fourth step was to produce a *paragraph* that linked the thinking sequence into prose. In a variety of guises, dependent on the purpose of the initial question asked and the wide range of phenomena that could be explored, the thinking frame provided a flexible pedagogic teaching tool that was widely used in the south of England. It

[1] Sonic Gun is a circular flexible membrane stretched over a circular frame is drawn back at its center and released. The force produces a shock wave in the air that is detectable at a short distance.

enabled all the ways of learning about models and modelling to be explored, up and including 'learning to model *de novo*'.

3. Providing a basis for evidence of learning.

Students in schools in England and Wales are required to demonstrate their learning in a series of public examinations. Of the types of written questions that they are asked, the most challenging is the 'short answer', where a series of sentences have to be composed by the student. This has been found to be very challenging to students. In the light of this experience, the Cams Hill Science Consortium developed a 'Literacy Ladder' to help teachers integrate writing skills into their use of the 'Thinking Frame' (see Fig. 10.6.).

When the sequence of activities in the ladder was undertaken with classes jointly with colleagues from the corresponding English Department, the outcome was found to be most successful: students could express themselves both more concisely and precisely.

4. Providing a scheme of work for the National Curriculum.

As the substantially nature of the revised National Curriculum became apparent during the 2010s, the Consortium was concerned that teachers would devalue their own extensive professional experience and instead purchase new published textbook schemes, some of which might have been inadequately field-tested. So it was

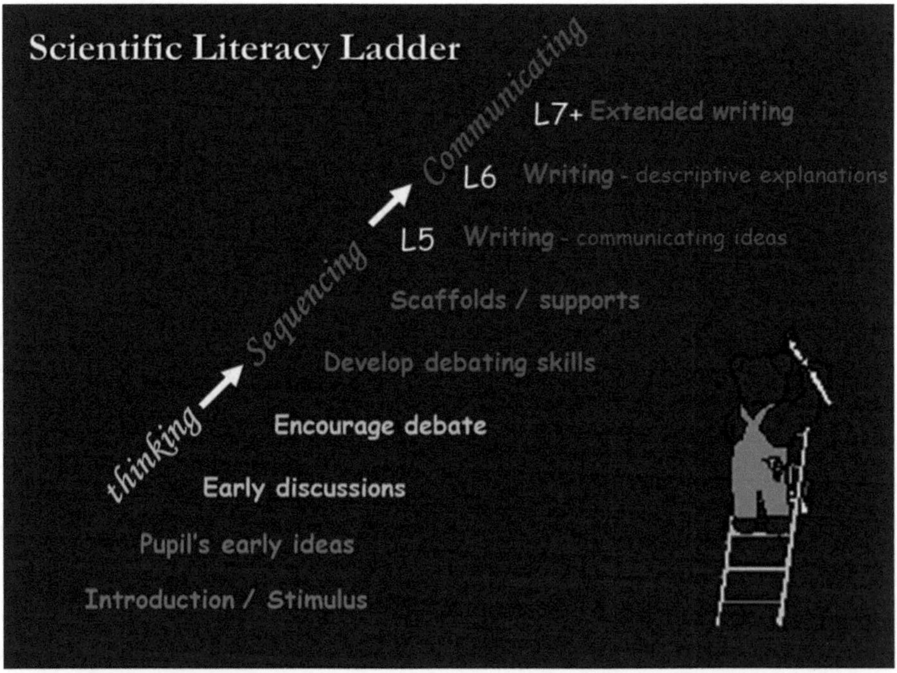

Fig. 10.6 The literacy ladder (Newberry & Cams Hill Science Consortium, 2007)

decided, perhaps as a culmination of the work of the Consortium, to produce a model of the curriculum that, whilst faithfully adhered to the mandatory requirements of the National Curriculum, also enabled teachers to develop lesson plans that valued their professional experience.

The basis for the model curriculum was the *topic overview*. A generic structure of a topic overview is given in Fig. 10.7.

The key aspects of a topic overview were:

- The topic title.
- The background colour allocated of the theme in the National Curriculum to which the topic belonged. The themes were: Survival (animals); Survival (plants); Changes; Forces; and Materials. In each of the 6 years of the primary school, students would study 6 topic overviews. In each of the 3 years of the junior high school, students would study 13 or 14 topic overviews. By junior high school the names of the key themes were updated to be called: Life & Survival; Energy; Forces; and Particles.
- The 'key vocabulary' would list both words that had been previously listed and the new words associated with the particular topic overview.
- The key concept that underlay the topic being studied.
- The other 'bubbles' indicate the facts about specific phenomena that are to be studied.

Examples are given in Figs. 10.8 and 10.9.

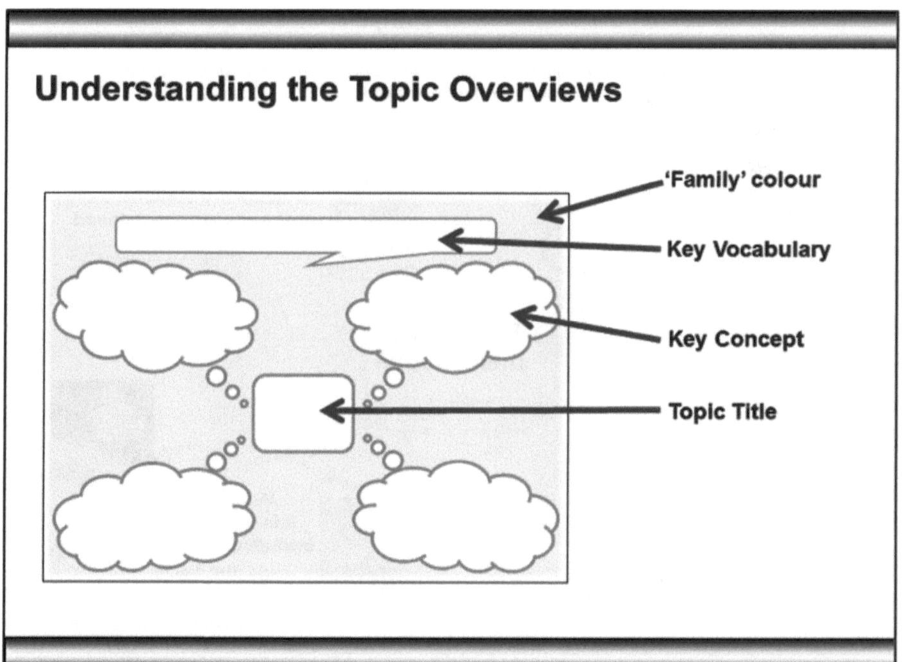

Fig. 10.7 The structure of a topic overview (Newberry & Cams Hill Science Consortium, 2014)

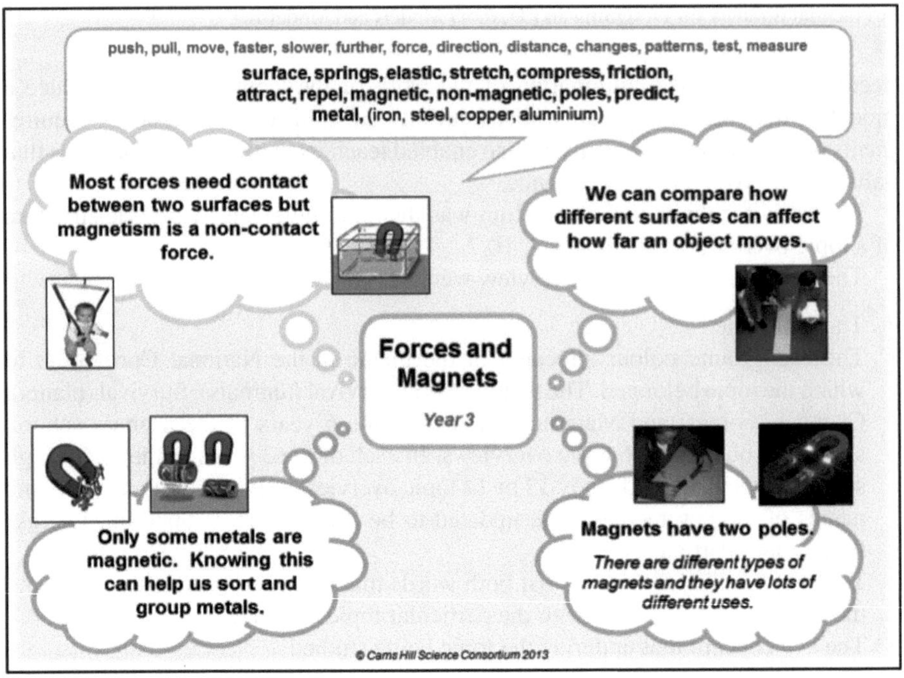

Fig. 10.8 A topic overview from a Year 3 forces topic (Newberry & Cams Hill Science Consortium, 2014)

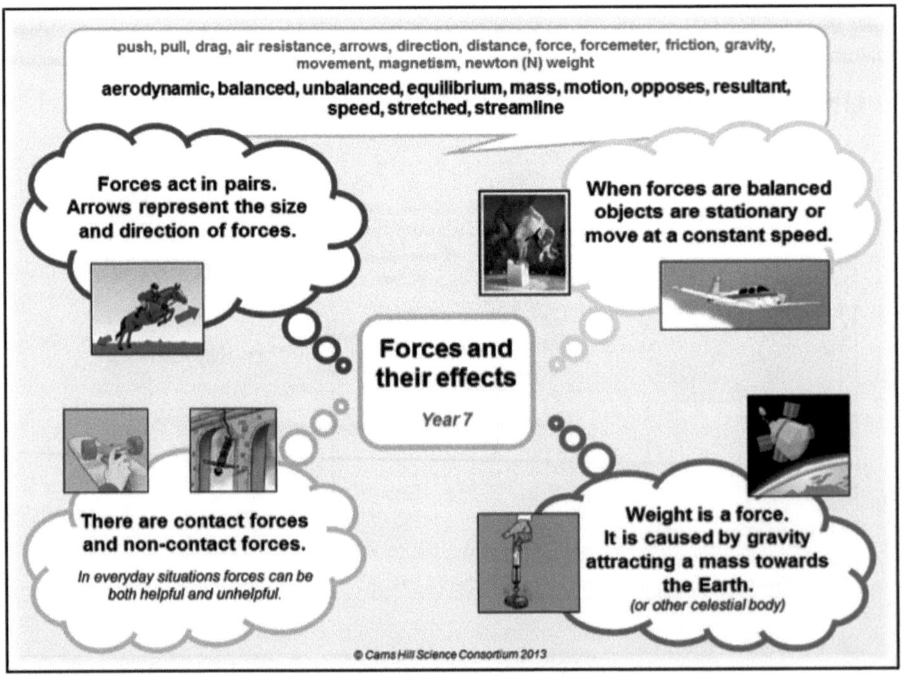

Fig. 10.9 A topic overview from a Year 7 forces topic (Newberry & Cams Hill Science Consortium, 2014)

Further details of the Cams Hill Science Consortium can be found on http://www.sciencepathways.co.uk, where copies of these resources may be purchased. At the time of writing (June 2015), over 150 primary and secondary schools have adopted the scheme.

Addressing the Challenges of Implementing an LP on Modelling

Whilst a broad framework for the design and implementation of an LP on modelling is a necessary first step in its implementation, teachers must take practical steps to realise its potential in the particular realities of the classroom, laboratory, field centre or other learning environment. Several such steps are discussed below.

Gaining Access to Phenomena

The phenomena that can sustain an LP on modelling must, of course, be able to support both the acquisition of all the elements of both the knowledge of modelling and of the development of the practice of modelling. It also must, in the realities of a school science curriculum, lead to the attainment of a high-level understanding of some of the core content concepts required by that curriculum. When should this be attempted with students and how might it be organised?

Unless a step-wise 'cognitive capability' model of conceptual development is adopted by teachers, there seems every reason why attempts to implement an LP on modelling might be undertaken with students of any age. A few detailed examples have been published on the use on such work with younger pupils. One example is that of framing an evidently successful enquiry into 'the action of a solar still' by Kenyon, Schwarz, and Hug (2008) with Grade 5 classes. These authors also give sketch outlines of LPs for modelling for such students built around: the life cycle of insects, electrical circuits, condensation, human sight, how smell works. However, in a review of papers that have appeared on modelling- oriented assessment in the period 1980–2013, Namdar and Shen (2015) showed that only a very low proportion (1/3) of the students could grasp the purpose of the empirical work required when at high school level. However, as the notion of teaching towards an LP on modelling gains ground with teachers, textbook writers, and curriculum designers in the next few years, it must be expected that reports of classroom experience will begin to define the lower age-limit at which such work can realistically begin, if one is in fact necessary.

Another perspective on the issue of 'access' is: how should that be provided? The traditional valuation of 'direct empirical experience' in science education (see, for example, Hofstein & Lunetta, 1982) strongly suggests that this should be the preferred form. However, the pressure on curriculum time, the complexity of arranging such direct access, particularly for the long periods of time and for large numbers of students that modelling work entails, coupled with the wish to have that work be decisive and leading to clear cognitive gains, has led to the introduction of computer-based 'experience'. This has been coupled with guided instruction, especially in the examples authored by the Concord Consortium in the USA (The Concord Consortium, 2014).

If we have clear ideas both of what can reasonably be achieved with a particular group of pupils and of the means by which suitable access to appropriate experiences can be provided, what then might the characteristics of those experiences be? They must:

- address phenomena in which the students have an interest (or can be persuaded to find interesting!);
- be such that exemplar forms can be made available to students;
- depend on a few scientific concepts for its explanation.

However, the most important criterion must be that phenomena chosen must be capable of sustaining 'authentic modelling practices'.

Identifying and Modelling Phenomena that Are Candidates for 'Authenticity'

Authentic modelling practices in science are those that are characteristic of a group of workers in the field, who, as a result of addressing purposes in common, come to use a set of research skills based on the same pool of knowledge. Such practices, once identified with the help of the scientists, can be adapted for learning purposes, the major gain of this being that students will come to understand and use the epistemology that underpins them. Situations that are capable of doing so have to meet a number of criteria. They must be capable of:

- provoking interest in students because they relate directly to the impact of science on society;
- being enquired into by groups of students in an autonomous manner;
- involving the use of a coherent model of modelling;
- conveying the same subject knowledge as is required to understand the original scientific practice. This implies that details descriptors of that practice are needed;
- being enquired into safely using the equipment available in a school;
- not requiring too much curriculum time (after Prins, Bulte, van Driel, & Pilot, 2008).

In the 2008 paper, Prins et al. also identified three topics as meeting all their criteria: the modelling of microbiological contamination in food chains to predict food safety; the modelling of the water treatment process used to predict the quality of drinking water that can be produced from surface water; and the modelling of human exposure to and uptake of chemicals emitted by consumer products in order to predict the safety of consumer products. Prins et al. (2011) reported the evidently broadly successful implementation of the 'water treatment' case with ten groups of four 16–17 year-old (Grade 10 and 11) chemistry students in The Netherlands. However, the procedure occupied 8 lessons, each of 50 min, and suggests that such authentic modelling activities are not to be undertaken lightly, given their considerable expense of curriculum time. It is also the case that the implied effort and commitment needed to establish the scientific practice being taught authentically are considerable. There seems no way of short-circuiting access to the scientific knowledge and skills involved.

Ensuring That 'Transfer of Learning' Takes Place

Successful 'transfer of learning' means that knowledge which is learnt in one context can be employed in another context (Gick & Holyoak, 1980). Attaining such transfer when an LP on modelling is the focus of attention is especially important for two reasons, both of which have got to do with the use of time. First, developing students' full understanding of modelling will take a lot of curriculum time, so there is pressure to ensure that the investment is worthwhile. Second, whilst the LP on modelling is being addressed, students will have necessarily been learning concepts that are core components of the curriculum: it is important to show that the latter has been fully effective.

The issue of 'transfer' in relation to the knowledge and skills of modelling has not yet been the focus of many empirical studies. Bamberger and Davis (2013) differentiate between two idealised contexts in which transfer may take place. 'Near transfer' is when learning in situation A is similar to that required for situation B and the similarity between the two situations is readily apparent to students: they also call this 'transfer-in-situation'. 'Far transfer', on the other hand, is when learning in situation A is considerably different to that in needed for situation B and the relationship between the two is not readily apparent to students: they call this 'transfer-between-situations'. One would expect that 'transfer-between-situations', of the two, is much harder to achieve.

This same paper also rehearses the three levels of increased understanding attainable for each of the

> … four dimensions of the epistemic criteria that capture growth in students' performance and understanding of the practice (of an LP in modelling): (A) attention to abstraction and representation of the features of the model; (B) attention to clarity of communication and audience understanding; (C) attention to evidentiary support (for claims made); (D) attention to mechanistic and process-oriented versus illustrative/descriptive accounts (Bamberger & Davis, 2013, p. 216 as originally set out in; Schwarz et al., 2009).

From this, they set out three increasing levels of modelling performance that deal with each of:

> The *explanation* domain (which) refers to the extent to which the model (produced by a student) answers questions about *how* and *why* the scientific phenomenon happened. (…)

> The *comparative* domain refers to the extent to which the model compares the two situations. (…) The *abstraction* domain refers to which aspects of the model include elements that are inaccessible to our eyes. (…) The *labelling* domain refers to the extent to which the model includes a key and labels of the model's elements (Bamberger & Davis, 2013, p. 223).

The Bamberger and Davis (2013) study drew on standard four-level models of the understanding of 'the particle nature of matter' and of 'friction', and was conducted with Grade 6 students. They were taught an LP module on 'the particle nature of matter' (PNM) as manifest in 'smell', taking tests both before and after this experience of their understanding of 'smell','evaporation', and 'friction'. The results showed that for:

- the same topic as that of the teaching ('smell'), the students improved both their understanding of the PNM and their modelling performance. In respect of the latter, students showed improvement in the 'explanation' and 'communication' domains but less so in the 'comparativeness' domains and not at all in the 'abstraction' domains;
- a question concerning 'evaporation', a topic that is in a near-content transfer-in-situation to 'smell', the students showed significant improvement in their understanding of PNM but only in that of the 'explanation' domain of their model. In short, their modelling capabilities of explanation increased as long as the content was readily perceived as familiar;
- a question concerning 'friction', a far-content transfer-in-situation topic, the students did not improve their content knowledge, but did improve their modelling performances in the 'explanation', 'abstraction' and 'labelling' domains.

The results of this study show that, when an LP on modelling is the focus of teaching, students can transfer that knowledge to new content areas, provided that this calls for 'near transfer' (also known as 'transfer-in-situation'), but that this transfer was not necessarily reflected in their knowledge of the content of the new area. These results also suggest that what students perceive to be the focus of the teaching is all important in deciding what they learn. In practice, therefore, teachers have to strive to give a balanced emphasis on 'modelling skills' and 'content knowledge acquisition' for transfer to be achieved in even 'near transfer situations'. Another implication that might then be drawn, by extrapolation from these results, is that 'far transfer' may require even more prolonged and focused attention when modelling is used in order to achieve an improved content knowledge.

Establishing LPs in Modelling

Whilst some progress has been made in identifying the characteristics of LPs in modelling, they will only be widely facilitated if:

- a 'library' of phenomena that are readily capable of providing opportunities for work on 'models and modelling' is established with the skills and knowledge of modelling entailed for each point of entry having been identified;

- the 'entry' and 'exit' characteristics of attainment for students of different ages are established for each phenomenon;
- the intermediate steps in progression are established, recognising that this will be far easier in terms of 'knowledge' than for 'skills';
- detailed strategies for teaching these ideas are established.

This list implies a major effort of research and development by the science education community. There is one more theme that cannot be overlooked: assessment.

The Assessment of Progression Towards Competence in Modelling

If students are to achieve meta-competence in models and modelling, then a major factor must be their engagement in 'self-assessment' (James, Black, McCormick, & Wiliam, 2006). This will contribute to the formative evaluation conducted by the teacher, which is defined as:

> Evaluation conducted while a creative process is under way, designed and used to promote growth and improvement in a students' performance or in a program development (Gullickson, 2002)

and no doubt contained within an overall strategy focused on inquiry-based science education (Csapo, 2014).

Despite the centrality of this activity, a recent review by Nicolau and Constantinou (2014) found that only 23 published papers about it were substantial, concerned with school-age students, and were empirically-based. It seems that, in general, data is collected by use of interviews, open- and closed-questionnaires, video, and concept mapping. The data collected was concerned with modelling practices, the products of modelling (i.e. models), the acquisition of meta-knowledge, and the cognitive processes involved. The authors observe that "in the reviewed papers, modelling competence was neither defined nor assessed in a unified manner" (p. 71).

The theme of the assessment of models and modelling, both for formative and summative purposes, does then require considerable additional work, not least because their inclusion in the 'Next Generation National Standards for Science' for the USA, which is likely to be very influential at world-level, is clearly signalled (NGSS Lead States, 2013).

References

Abell, S. K., & Roth, M. (1995). Reflections on a fifth-grade life science lesson: Making sense of children's understanding of scientific models. *International Journal of Science Education, 17*(1), 59–74.

Arnold, M., & Millar, R. (1996). Learning the scientific 'story': A case study in the teaching and learning of elementary thermodynamics. *Science Education, 80*(3), 249–281.

Aubusson, P. J., Harrison, A. G., & Ritchie, S. M. (2006). *Metaphor and analogy in science education*. Dordrecht, The Netherlands: Springer.

Aubusson, P. J., Treagust, D. F., & Harrison, A. G. (2009). Learning and teaching with analogies and metaphors. In S. M. Ritchie (Ed.), *The world of science education: Handbook of research in Australasia* (pp. 199–216). Rotterdam, The Netherlands: Springer.

Bamberger, Y. M., & Davis, E. A. (2013). Middle-school science students' scientific modelling performances across content areas and within a learning progression. *International Journal of Science Education, 35*(2), 213–238.

Barab, S. A., Hay, K. E., Barnett, M., & Keating, T. (2000). Virtual solar system project: Building understanding through model building. *Journal of Research in Science Teaching, 37*(7), 719–756.

Bloom, B. S. (1956). *A taxonomy of educational objectives*. New York, NY: David McKay.

Chittleborough, G. D., Treagust, D. F., Mamiala, T. L., & Mocerino, M. (2005). Students' perceptions of the role of models in the process of science and in the process of learning. *Research in Science & Technological Education, 23*(2), 195–212.

Corcoran, T., & Silander, M. (2009). Instruction in high schools: The evidence and the challenge. *The Future of Children: America's High Schools, 19*, 157–183. Retrieved from http://www.futureofchildren.org

Csapo, B. (2014). *Report on a stretagy for the assessment of skills and competences suitable for IBSE*. Dublin, Ireland: Dublin City University.

Deng, F., Chen, D.-T., Tsai, C.-C., & Chai, C. S. (2011). Students' views of the nature of science: A critical review of research. *Science Education, 95*(6), 961–999.

Driver, R., Leach, J., Millar, R., & Scott, P. (1996). *Young people's images of science*. Buckingham, UK/Philadelphia, PA: Open University Press.

Duschl, R., Maeng, S., & Sezen, A. (2011). Learning progressions and teaching sequences: A review and analysis. *Studies in Science Education, 47*(2), 123–182.

Eilam, B. (2012). *Teaching, learning, and visual literacy*. Cambridge, UK: Cambridge University Press.

Erduran, S., & Jiménez-Aleixandre, M. P. (2008). *Argumentation in science education: Perspectives from classroom-based research*. Dordrecht, The Netherlands: Springer.

Ford, M. J. (2015). Learning progressions and progress: An introduction to our focus on learning progression. *Science Education, 99*(3), 407–409.

Garcia-Mila, M., & Andersen, C. (2008). Cognitive foundations of learning argumentation. In S. Erduran & M. P. Jiménez-Aleixandre (Eds.), *Argumentation in science education: Perspectivs from classroom-based research* (pp. 29–45). Dordrecht, The Netherlands: Springer.

Gick, M. L., & Holyoak, K. J. (1980). Analogical problem solving. *Cognitive Psychology, 12*(3), 306–355.

Gilbert, J. K., & Eilam, B. (2014). Developing science teachers' representational competence and its impact on their teaching. In B. Eilam & J. K. Gilbert (Eds.), *Science teachers' use of visual representations* (pp. 315–329). Dordrecht, The Netherlands: Springer.

Gilbert, J. K., Justi, R., & Queiroz, A. S. (2010). The use of a Model of Modelling to develop visualization during the learning of ionic bonding. In M. F. Taşar & G. Çakmakcı (Eds.), *Contemporary science education research: International perspectives* (pp. 43–51). Ankara, Turkey: Pegem Akademi.

Gilbert, J. K., & Watts, D. M. (1983). Conceptions, misconceptions, and alternative conceptions: Changing perspectives in science education. *Studies in Science Education, 10*(1), 61–98.

Gobert, J. D., & Pallant, A. (2004). Fostering students' epistemologies of models via authentic model-based tasks. *Journal of Science Education and Technology, 13*(1), 7–22.

Goswami, U. (1992). *Analogical reasoning in children*. Hillsdale, NJ: Erlbaum.

Gotwals, A. W., & Alonzo, A. C. (2012). Introduction: Leaping into learning progressions in science. In A. C. Alonzo & A. W. Gotwals (Eds.), *Learning progressions in science – current challenges and future directions* (pp. 3–12). Rotterdam, The Netherlands: Sense.

Grosslight, L., Unger, C., Jay, E., & Smith, C. L. (1991). Understanding models and their use in science: Conceptions of middle and high school students and experts. *Journal of Research in Science Teaching, 28*(9), 799–822.

Gullickson, A. R. (Ed.). (2002). *The student evaluation standards: How to improve evaluations of students*. Thousand Oaks, CA: Corwin Press.

Halbern, D., & Collaer, M. (2005). Sex difference in visuospatial abilities: More than meets the eye. In P. Shah & A. Miyake (Eds.), *The Cambridge handbook of visuospatial thinking* (pp. 170–212). Cambridge, UK: Cambridge University Press.

Halloun, I. A. (1998). Schematic concepts fro schematic models of the real world: The Newtonian concept of force. *Science Education, 82*(2), 239–263.

Harrison, A. G., & Treagust, D. F. (2006). Teaching and learning with analogies: Friend or foe? In P. Aubusson, A. Harrison, & S. M. Ritchie (Eds.), *Metaphor and analogy in science education* (pp. 11–24). Dordrecht, The Netherlands: Springer.

Hegarty, M., & Waller, D. (2005). Individual differences in spatial abilities. In P. Shah & A. Miyake (Eds.), *The Cambridge handbook of visuospatial thinking* (pp. 121–169). Cambridge, UK: Cambridge University Press.

Hofstein, A., & Lunetta, V. (1982). The role of the laboratory in science teaching: Neglect aspects of research. *Review of Educational Research, 52*(2), 201–217.

Hogan, K. (2000). Exploring a process view of students' knowledge about the nature of science. *Science Education, 84*(1), 51–70.

James, M., Black, P., McCormick, R., & Wiliam, D. (2006). Learning how to learn, in classrooms, schools and networks: Aims, design and analysis. *Research Papers in Education, 21*(2), 101–118.

Justi, R., Gilbert, J. K., & Ferreira, P. F. M. (2009). The application of a 'model of modelling' to illustrate the importance of metavisualization in respect of the three levels of representation. In J. K. Gilbert & D. F. Treagust (Eds.), *Multiple representations in chemical education* (pp. 285–307). Dordrecht, The Netherlands: Springer.

Kang, S., Scharmann, L. C., & Noh, T. (2005). Examining students' views on the nature of science: Results from Korean 6th, 8th, and 10th graders. *Science Education, 89*(2), 314–334.

Kenyon, L., Schwarz, C. V., & Hug, B. (2008). The benefits of scientific modeling. *Science and Children, 46*(2), 40–44.

Khishfe, R. (2008). The development of seventh graders' views of nature of science. *Journal of Research in Science Teaching, 45*(4), 470–496.

Leach, J., Hind, A., & Ryder, J. (2003). Designing and evaluating short teaching interventions about the epistemology of science in high school classrooms. *Science Education, 87*(6), 831–848.

Lehrer, R., & Schauble, L. (2012). Seeding evolutionary thinking by engaging children in modeling its foundations. *Science Education, 96*(4), 701–724.

Louca, L. T., Zacharia, Z. C., & Constantinou, C. P. (2011). In quest of productive modeling-based learning discourse in elementary school science. *Journal of Research in Science Teaching, 48*(8), 919–951.

Luxford, C. J., & Bretz, S. L. (2013). Moving beyond definitions: What student-generated models reveal about their understanding of covalent bonding and ionic bonding. *Chemistry Education Research and Practice, 14*(2), 214–222.

Maia, P. F., & Justi, R. (2009). Learning of chemical equilibrium through modelling-based teaching. *International Journal of Science Education, 31*(5), 603–630.

Mendonça, P. C. C., & Justi, R. (2013). The relationships between modelling and argumentation from the perspective of the Model of Modelling diagram. *International Journal of Science Education, 35*(14), 2007–2034.

Mirza, N. M., & Perret-Clermont, A.-N. (2009). *Argumentation and education – theoretical foundations and practices*. Dordrecht, The Netherlands: Springer.

Mozzer, N. B., & Justi, R. (2012). Students' pre- and post-teaching analogical reasoning when they draw their analogies. *International Journal of Science Education, 34*(3), 429–458.

Namdar, B., & Shen, J. (2015). Modeling-oriented assessment in K-12 science education: A synthesis of research from 1980 to 2013 and new directions. *International Journal of Science Education, 37*(7), 993–1023.

Newberry, M. (2006). *Using the 'Thinking Frames' approach to improve pupil engagement and attainment in science*. Paper presented at the teacher research conference, London, UK.

Newberry, M., & Cams Hill Science Consortium. (2007). *Thinking frame*. Retrieved from http://www.thinkingframe.com

Newberry, M., & Cams Hill Science Consortium. (2014). *Science pathways*. Retrieved from http://www.sciencepathways.co.uk

Newberry, M., & Gilbert, J. K. (2007). Bringing learners and scientific expertise together. In K. S. Taber (Ed.), *Science education for gifted learners* (pp. 212–217). London, UK: Routledge.

Newberry, M., Grevatt, A., & Gilbert, J. K. (2009). *Science pathways for year 7*. Stevenage, UK: Badger.

Newcombe, N. S., & Learmonth, A. E. (2005). Development of spatial competence. In P. Shah & A. Miyake (Eds.), *The Cambridge handbook of visuospatial thinking* (pp. 213–256). Cambridge, UK: Cambridge University Press.

NGSS Lead States. (2013). *Next generation science standards: For states, by states*. Washington, DC: National Academies Press.

Nicolau, C. T., & Constantinou, C. P. (2014). Assessment of modelling competence: A systematic review and synthesis of empirical research. *Educational Research Review, 13*, 52–73.

Niss, M. (2009). Metamodelling messages conveyed in five statistical mechanical textbooks from 1936 to 2001. *International Journal of Science Education, 31*(5), 697–719.

Passmore, C. M., & Stewart, J. (2002). A modeling approach to teaching evolutionary biology in high schools. *Journal of Research in Science Teaching, 39*(3), 185–204.

Press, H. (1995). *The little book of scientific experiments*. New York, NY: Strirling.

Prins, G. T., Bulte, A. M. W., & Pilot, A. (2011). Evaluation of a design principle for fostering students' epistemological views on models and modelling using authentic practices as contexts for learning in chemistry education. *International Journal of Science Education, 33*(11), 1539–1569.

Prins, G. T., Bulte, A. M. W., van Driel, J., & Pilot, A. (2008). Selection of authentic modelling practices as contexts for chemistry education. *International Journal of Science Education, 30*(14), 1867–1890.

Redfors, A., & Ryder, J. (2001). University physics students' use of models in explanations of phenomena involving interaction between metals and electromagnetic radiation. *International Journal of Science Education, 23*(12), 1283–1301.

Ryder, J., Leach, J., & Driver, R. (1999). Undergraduate science students' images of science. *Journal of Research in Science Teaching, 36*(2), 201–219.

Schwarz, C. V., Reiser, B. J., Davis, E. A., Kenyon, L., Achér, A., Fortus, D., … Krajcik, J. (2009). Developing a learning progression for scientific modeling: Making scientific modeling accessible and meaningful for learners. *Journal of Research in Science Teaching, 46*(6), 632–654.

Shulman, L. S. (1987). Knowledge and teaching: Foundations of the new reform. *Harvard Educational Review, 51*(1), 1–22.

Sins, P. H. M., Savalsberg, E. R., & van Joolingen, W. (2005). The difficult process of scientific modelling: An analysis of novice's reasoning during computer-based modelling. *International Journal of Science Education, 27*(14), 1695–1721.

The Concord Consortium. (2014). *Showcase*. Retrieved from http://mw.concord.org/modeler/showcase

Treagust, D. F., Harrison, A., & Venville, G. J. (1998). Teaching science effectively with analogies: An approach for preservice and inservice teacher education. *Journal of Science Teacher Education, 9*(2), 85–101.

Chapter 11
Educating Teachers to Facilitate Modelling-Based Teaching

Abstract Successful modelling-based teaching makes specific and complex demands on the knowledge and skills of science teachers. Of the various categories of knowledge and skills involved, pedagogic content knowledge (PCK) has been the hardest to define precisely in respect of school-level education generally. In the relative absence until recently of research into the PCK involved specifically in MBT, and whilst we only have 'best practice' criteria for the conduct of science teacher education in general, little is known to date about how the PCK of MBT can be developed. In this chapter we have attempted to identify the PCK involved in MBT and have speculated on how that knowledge evolves in the light of an established model of 'good practice' over the passage of time and the aggregation of relevant professional experience. Most importantly, we have identified strategies that may be adopted by teachers within the practice of MBT if the requisite knowledge and skills are to become widely available.

Rationale for Teachers' Education for Modelling-Based Teaching

When discussing modelling-based teaching (MBT) in Chap. 4, we emphasised the five approaches to learning about models and modelling that we had previously identified in the literature: *learning curricular models*; *learning to use models*; *learning to revise models*; *learning to reconstruct models*; *learning to construct a model de novo*. In that chapter, we also classify the three initial approaches as *model-based teaching* and the last two as *modelling-based teaching* (Fig. 4.1) – those that comprise our main interest in this book. In this chapter we emphasise that, independently of the approach adopted, teachers play a key role in supporting students' learning. After all, teachers are those who mainly decide which curricular model will be introduced to students; which teaching models will be used to support students understanding of a given curricular model; in which new contexts students should try to use a given model; which distinct purposes a given model should fulfil; in which distinct mode of representation students should express a given model; with which experiences students will be provided; and which questions they will be asked in order to create, test, and evaluate a model. Additionally, teachers are those

J.K. Gilbert, R. Justi, *Modelling-based Teaching in Science Education*, Models and Modeling in Science Education 9, DOI 10.1007/978-3-319-29039-3_11

who mainly motivate students to engage in modelling-based activities, who provide emotional support for those experiencing difficulties in MBT activities, and who encourage them to express and defend their ideas. Therefore, engaging students in model- or modelling-based teaching activities may place a high demand on teachers (Justi & Gilbert, 2002a; Kenyon, Davis, & Hug, 2011; Kloser, 2014). Equally importantly, if and how students are engaged in model- or modelling-based teaching activities is directly influenced by teachers' knowledge of, and beliefs in, such teaching approaches (Windschitl & Thompson, 2006).

Being more specific, the literature identifies a series of types of knowledge and skills that teachers should have in order to plan and conduct MBT. They should understand: the meaning, the nature, and the main purposes of models in science; the meaning and purposes of each of the stages of modelling; the meaning and purposes of engaging students in MBT activities; the ways MBT can be integrated into classroom practices; and how to conduct MBT in regular classes. They should also have had experience of: modelling practices; evaluating how well the basic features of a given model are clearly represented in an expressed model; recognising the strengths and limitations of each model in a series of multiple competing ones; identifying the conditions for using MBT in regular classrooms; the main strategies that may be integrated into such situations and how to use them in order to support students learning (Davis et al., 2008; Justi, 2009).

Unfortunately, previous attempts to characterise such knowledge and skills showed that, in general, teachers are not competent in this area, and their classroom practices rarely include modelling activities. For instance, van Driel and Verloop (1999) used open and Likert-type scale questionnaires to investigate the knowledge of a sample of 86 Dutch experienced science teachers about models (mainly the nature and the roles of models in science) and modelling (only in terms of general ideas about the design and revision of models). Although the results indicated that most of the teachers viewed models as simplified representations of reality, they also showed that teachers neither use regular criteria to classify a series of examples as models nor recognise that making predictions is a relevant function of scientific models. Such evidence supported the authors' conclusion that the teachers' knowledge about models and modelling was limited.

In a following study, these same authors investigated the use of teaching activities related to models and modelling (van Driel & Verloop, 2002). This study was conducted in two phases. Initially, seven Dutch experienced teachers were interviewed, aiming at identifying general characteristics of their practical knowledge, that were then emphasised in the items of a Likert-type scale questionnaire answered by 74 experienced science teachers. In both phases, the authors identified two subgroups of teachers: those who declared using many types of teaching activities focused on models and modelling, and those who reported using few of them. Additionally, there seemed to be no clear relationship between the self-reported use of such teaching activities and the teachers' knowledge of their students' views on models. Due to the level of experience of the sample (media of 17.5 years), this lack of integration between the teachers' use of specific instructional strategies and their knowledge of students' conceptions and difficulties concerning models was viewed as a surprise.

In another broad-based study, 39 Brazilian (primary, secondary, university, and pre-service) science teachers were interviewed in order to characterise their knowledge about models, modelling, and their use in teaching, as well as their actions concerning the introduction of models and modelling in all levels of science teaching (Justi & Gilbert, 2002a, 2002b, 2003). An approach to analysing the teachers' notion of models was used, based on seven aspects (the *nature* of a model, the *use* to which it can be put, the *entities* of which it consist, its relative *uniqueness*, the *time* span over which it is used, its status in the making of *predictions*, and the basis for the *accreditation* of its existence and use), and a series of categories of meaning for all of them was proposed. Several distinct views emerged from the data, although no clear profiles of understanding for individuals were identified – which was interpreted as evidence that most teachers might not hold consistent ontological and epistemological views on models.

The teachers' notions of modelling was also found to be unsatisfactory since, for instance, most of them did not (i) recognise modelling as an inherent and dynamic process of knowledge production; (ii) identify the role of distinct modes of representation (or even recognise some of them as possible ways to express a model); and (iii) acknowledge the relevance of considering the scope and limitations of a given model in the process. Despite the inconsistencies in their ideas about models and modelling, most of the teachers showed themselves to be aware of the value of models in the learning of science. However, they did not show an awareness of the role of models in the learning *about* science. In fact, none of them had even expressed any ideas concerning learning *about* science. For them, science teaching was restricted to teaching (the transmission of) the content of science. This explains, for instance, why almost all of those who assumed that models might be used in teaching understood them as teaching models, that is, as aids that may support students' visualisation and/or understanding of the content under discussion.

Finally, even recognising the limitations of analysing teachers' declarative knowledge, the authors decided to explore possible relationships between the teachers' general ideas on the use of models and modelling in teaching and what they said when the interviewer tried to contextualise such ideas in their own teaching practices. This was done by analysing the links between their beliefs about their students' views on models, the value they place on involving students in modelling activities, and how they respond to the outcomes of such activities. Most of the teachers (from all the sub-sets) were not worried about students' understanding of models or did not pay attention to what their students say. On the other hand, 59 % of the teachers (including almost all primary level ones) positively valued the engagement of students with modelling activities. However, from their responses it become clear that most of them understood "modelling activities" as those in which students would have to build a concrete model for some previously known entity. Other 21 % of the teachers said that sometimes they 'discussed' their students' models (generally only by asking them to explain their models), but that, at the end, those models were ignored and the curricular models were imposed. Assuming that among this group of teachers there were a lot of those who worked at the university level, that is, who were involved in the education of future secondary education teachers, the fact that MBT was not a common practice causes no surprise. Additionally, the fact that many researchers, from different countries, have been promoting initiatives to support teachers' professional development in this area (as it is discussed later in this chapter) corroborates our conclusion that such a vague, limited and inconsistent view on MBT is not exclusive to the Brazilian teachers who participated in this study.

Therefore, what we discussed in this section constitutes a good rationale for teachers' education in this area. Ultimately, if teachers are those responsible for introducing and conduction MBT in regular science classes in all educational levels, and if they do not have the knowledge and skills necessary to support such actions, it seems crucial to improve their education. The analysis of the literature made it evident that all researchers who have been conducting studies in this area recently have been motivated to do so by this conclusion. Such studies have been mainly conducted with pre-service teachers, probably as part of the attempts to support their understandings of approaches and strategies to teaching as distinct from the traditional knowledge-transmission one used during their own education, that is, before they start working regularly as a teacher. Before discussing such studies, and also in order to support a clear organisation of the discussion, in the next section we shift our focus to the general literature on teachers' knowledge.

Teachers' Knowledge

A Complex Set of Categories of Knowledge

The knowledge and skills that someone has to develop in order to become a good teacher have been investigated for many years. Although it is out of the scope of this chapter to discuss the bulk of the literature on this area, it is necessary to present and comment on some ideas that have had a great influence on the studies about science teachers' knowledge (and, in particular, their knowledge about modelling).

Lee Shulman's presidential address at the 1985 annual meeting of the American Educational Research Association (Shulman, 1986) was a landmark in research about teachers' knowledge. It was soon followed by another publication (Shulman, 1987) that completed and expanded some ideas that had attracted special attention among the educational research community leading to their acceptance as the basis for most of the research conducted in the area since then. In the first publication, Shulman reframed the study of teachers' knowledge by emphasising what he called a *missing paradigm*: the role of content in teaching. When doing so, he distinguished three categories of teachers' content knowledge:

- *subject matter content knowledge*, the understanding of the structure of the subject matter, that is, "the variety of ways in which the basic concepts and principles of the discipline are organized to incorporate its facts" and "the set of ways in which truth or falsehood, validity or invalidity, are established" (Shulman, 1986, p. 9). This means that teachers would not only be expected to define the accepted ideas from a given domain, but also to be able to present and defend a clear rationale for the teaching of each idea in terms of both the discipline and the general knowledge a student has to acquire to become scientific literate;
- *pedagogical content knowledge*, that is, "the subject matter knowledge *for* teaching" or, in different words, "the ways of representing and formulating the subject that make it comprehensible to others" and "the understanding of what makes the learning of specific topics easy or difficult" (Shulman, 1986, p. 9); and
- *curricular knowledge*, that is, "the full range of programs designed for the teaching of particular subjects at a given level" and "the variety of instructional materials available in relation to those programs" (Shulman, 1986, p. 10).

Four more categories were added in the later publication: *general pedagogical knowledge*; *knowledge of learners and their characteristics*; *knowledge of educational contexts*; and *knowledge of educational ends, purposes and values and their philosophical and historical grounds* (Shulman, 1987). Shulman's categories of teachers' knowledge were modified and reorganised later, which resulted in the proposition of distinct models, the most used being those proposed by Grossman (1990) and (Magnusson, Krajcik, & Borko, 1999) (the latter considering specifically the context of science teaching).

Independently of their particularities, all the subsequent models maintained the separation between *subject matter knowledge* (SMK) and *pedagogical content knowledge* (PCK), whose distinctions and characterisations have been focused on in a significant number of publications (for instance, Abell, 2008; Barnett & Hodson, 2001; Berry, Loughran, & van Driel, 2008; Friedrichsen, van Driel, & Abell, 2010; Gess-Newsome & Lederman, 1999; Kind, 2009; Loughran, 2006; Loughran, Berry, & Mulhall, 2006; Nilsson, 2008; Park & Oliver, 2008; van Driel, Meirink, van Veen, & Zwart, 2012). There has also been a consensus in the literature about the outstanding importance of the construct PCK. After all, as emphasised by Shulman in an interview conducted more than 20 years after the proposition of the construct, "just knowing the content well was really important, just knowing general pedagogy was really important and yet when you added the two together, you didn't get the teacher" (Shulman, quoted in Berry et al., 2008, p. 1274). So, it seems that two attributes of the construct PCK contribute to make it valuable as the core element in the knowledge base of teaching: its foundation of being an amalgam of essential distinct knowledge in teachers' profession; and its nature of being influenced by the subject matter (that is, of recognising that the teaching of any particular content does not occur from teachers' general pedagogical knowledge (Barnett & Hodson, 2001)).

On the other hand, the increasing interesting on the construct has also resulted in a growing number of particular views and definitions for it, thus making difficult both to find out exactly which elements comprise it, and thus how to use it to support good teaching practices (Kind, 2009; Loughran, Mulhall, & Berry, 2004). Such particular views were classified as *integrative* or *transformative* PCK (Gess-Newsome, 1999). Gess-Newsome has also drawn an analogy that helps us to understand the difference between them. According to her, those who understand PCK from an integrative perspective view it as teachers' knowledge as a whole comprised by subject matter knowledge, pedagogy, and context – like a chemical mixture. On the other hand, those who understand PCK from a transformative perspective visualise it as the product of a chemical change, that is, as the result of the transformation of subject matter, pedagogical, and contextual knowledge (the reactants) into a distinct type of knowledge. We agree that PCK results from a kind of transformation of other teachers' knowledge, but we would like to extend the original analogy. We view PCK not as the product of a chemical reaction, but as one of the constituents of a chemical equilibrium. In this sense, it is not only distinct from all other types of knowledge, but also influences, or contributes to the transformation, of them (Grossman, 1990; Sperandeo-Mineo, Fazio, & Tarantino, 2005). This outcome is a consequence of the reality that teacher's knowledge development is a continuous and dynamic process in which the teacher does not only have to understand a given topic in order to plan and conduct its teaching in a given context. From the planning and the conducting stages of the teaching, the teacher also continuously modifies his/her understandings of the topic (as well as other types of knowledge involved in the process). Thus, we propose to name this kind of bidirectional amalgamation as being *dual-transformative*.

It seems that the dual-transformative view is represented in the consensus model of teacher professional knowledge bases produced by a selected group of researchers who participated in a small working conference held in the USA in 2012. According to their consensus definition, a teacher's personal PCK is

> the knowledge of, reasoning behind, and planning for teaching a particular topic in a particular way for a particular purpose to particular students for enhanced student outcomes" and "the act of teaching a particular topic in a particular way for a particular purpose to particular students for enhanced student outcomes (Gess-Newsome, 2013, slide 14).

According to this consensus view, a teacher's PCK derives from (i) a set of professional knowledge bases: content knowledge, pedagogical knowledge, curricular knowledge, knowledge of students, and assessment knowledge; and (ii) a set of specific professional knowledge: knowledge of instructional strategies, content representations, student understandings, science practices and habits of mind related to a given topic. Returning to our amplified analogy, all these types of knowledge (including PCK) would participate in an equilibrium system placed in a given context, and in which the activation energy required for the occurrence of all reactions would be provided by the teacher's beliefs and orientations. As is the case in all chemical equilibriums, the transformations would never stop.

By understanding teachers' knowledge at this level of complexity, it becomes clear that to analyse and to foster its development are not simple processes. Thus, before analysing the studies focused on teachers' knowledge of modelling, and in order to scaffold some elements of our analysis, we briefly discuss some general guidelines for the development of teachers' knowledge.

The Development of Teachers' Knowledge – Some General Guidelines

Teacher education has always been one of the challenging enterprises faced by educational researchers and teacher educators. It seems plausible to think that the enterprise starts from the answer to a single question: what does it mean to educate a teacher? The answer reflects, however, the complexity inherent in the practice of teaching and in the meaning of 'being a teacher'. In the science education community, which tends to take into account students' ideas and to advocate that students should learn by a constructivist-based approach, many researchers adopt the same approach to teacher education. For them, to educate a teacher is to provide support for their active learning, to give them opportunities to build the necessary knowledge and to develop appropriate skills (Bell & Gilbert, 1996; Borko, 2004; Burbank & Kauchak, 2003; Capps & Crawford, 2013; Clarke & Hollingsworth, 2002; E.A. Davis, 2003; Reiser, 2013; van Driel & Berry, 2012 among others). From this approach, these and other researchers propose and/or use general guidelines for teacher education.

Following the conduct and analysis of data collected in a 3-year teacher development project in New Zealand, Bell and Gilbert (1996) suggest that there are three aspects to teacher development in general and which are therefore manifest in respect of any attempt to do so:

- *Social development.* This is the reconstruction and critical development by a teacher of (i) what it means to be a science teacher; and (ii) working with others, for example students, other teachers;
- *Personal development.* This is the construction, evaluation, and critical acceptance by an individual of an understanding of the socially reconstructed knowledge of what it means to be a science teacher;
- *Professional development.* This is the development of that repertoire of beliefs, knowledge, and skills that support the production of teaching activities, and enable the sense of being a science teacher to be exercised in everyday classroom practice.

These three elements develop over time and are manifest in relation as all the types of knowledge discussed in the previous sub-section are acquired and changed. In teachers' professional development programmes (PD), such elements have been explicitly addressed by the recognition that teachers may have alternative conceptions in respect of any type of knowledge (particularly the 'subject matter' and 'pedagogic content' ones) that must be focused on and be the focus of continuous professional development. This has been mainly done by:

- creating conditions for teachers to work with other teachers by discussing their ideas, actions, difficulties, and collectively and collaboratively renegotiating the meaning of being a science teacher. In other words, making them participate in a *community of practice* (Barnett & Hodson, 2001; Bell & Gilbert, 1996; Burbank & Kauchak, 2003; Clarke & Hollingsworth, 2002; Lin, Hong, Yang, & Lee, 2013; Putnam & Borko, 2000; Reiser, 2013);
- making teachers feel that they are taking part in a dynamic, non-linear (van Driel & Berry, 2012) and long-term (van Driel et al., 2012) developing process, and that in the learning process their ideas, experiences and beliefs are valued (Bell & Gilbert, 1996; K.S. Davis, 2003; Reiser, 2013). It is expected that to see themselves in such a supportive atmosphere may contribute to the reduction of any possible uncomfortable feelings (for instance, of being incompetent, inadequate, ignorant, inexperienced) which, in some sense, could influence teachers' motivation to change;
- providing teachers with information on new strategies and/or activities for teaching a particular topic, a good rationale for their uses, and examples of how to use them in regular classes. Such examples are particularly important because they contribute to situate teachers' learning experience in particular and common contexts (Putnam & Borko, 2000). This may help teachers to connect their own learning to their classrooms and to be more sure that their students would also learn from such strategies and/or activities (Bell & Gilbert, 1996; Capps & Crawford, 2013; K.S. Davis, 2003; Reiser, 2013);
- encouraging, and giving support to, teachers to produce and/or enact new teaching strategies and/or activities, as well as giving feedback on the outcomes of them (Bell & Gilbert, 1996; Putnam & Borko, 2000; van Driel & Berry, 2012);
- encouraging teachers to adopt the role of a critical inquirer in their classes (Bell & Gilbert, 1996; Burbank & Kauchak, 2003; Lin et al., 2013; Magnusson et al., 1999; Zembal-Saul, Blumenfeld, & Krajcik, 2000), something that derives from Schön's (1987, 1991) proposition of teacher as a *reflective practitioner*. In his view, when practising the profession, a teacher shows *knowledge-in-action* (teachers' knowledge acquired from previous experiences and cognitive activity that are consciously or unconsciously mobilised in teachers' actions), *reflection-in-action* (a critical practice, a looking at and thinking about current events, experiences, feelings, and actions during classrooms, resulting in the building of new understandings to inform their current actions), and *reflection-on-action* (reflection that occurs after the events, experiences, feelings, and actions have taken place, when teachers think about what they have done, thought or felt, and analyse which changes would have resulted in different outcomes in that situation). Therefore, as reflection is a crucial element in the promotion of teachers' learning (Calderhead & Gates, 1993), they can use it as a basis for their social,

personal, and professional development – as evidenced by convincing comments expressed by the teachers who participated in Bell and Gilbert's (1996) project.

Teachers' Knowledge About Modelling

Characterisation of Specific Types of Teachers' Knowledge About Models and Modelling

Among the most common types of teachers' knowledge, 'subject knowledge' and 'pedagogical content knowledge' have been those focused in studies of teachers' knowledge about models and modelling. The emphasis placed in the literature on the meaning of each of them has varied, whilst some studies do not even classify the knowledge investigated.

The literature has characterising *content knowledge about models* by analysing the main aspects identified in teachers' general ideas about them. It seems that the initial inspiration for investigating teachers' ideas on models was provided by the study of Grosslight, Unger, Jay, and Smith (1991), in which students' and experts' understandings of models were classified according to their epistemological increasing complexity into three general levels. This was so because the original questions in the interview that they used to collect data (or a slightly modified version of them) constituted the basis of the instruments (written questions and interviews) that were used to identify teachers' views about models in other studies (Justi & Gilbert, 2003; van Driel & Verloop, 1999). Due to the nature of the instruments, in both of these studies researchers analysed teachers' declarative knowledge, which was discussed in terms of:

- *characteristics* of scientific models, *goals and functions* of models in science, and *representational modes* of scientific models (van Driel & Verloop, 1999);
- the *nature* of a model, the *use* to which it can be put, the *entities* that can be modelled, its relative *uniqueness*, the *time* span over which it is used, its status in respect of the making of *prediction*, and the basis of *accreditation* for its existence and use (Justi & Gilbert, 2003).

As it was discussed in the initial section of this chapter, both studies show that teachers have a limited content knowledge about models. More recent studies have corroborated this conclusion by either investigating teachers' ideas about some of these aspects (Justi & van Driel, 2005c; Krell & Krüger, 2015) or taking it for granted that most teachers need support to develop their knowledge about models (Kenyon et al., 2011; Windschitl & Thompson, 2006).

As discussed in Chap. 3, modelling is a complex process comprised of a series of sub-process that involve many elements. However, it seems that when it comes to discussing teachers' *content knowledge on modelling*, the literature has assumed a general view of modelling as a process of producing and using models, or even some specific ideas related to, or included in, such a process. For instance, van Driel and Verloop's (1999) questions on modelling emphasised general aspects related do the design and development of models (for instance, the role of creativity in the development of models, the possibility of building a model by modifying another one). On the other hand, rather than asking directly what teachers thought about modelling, Justi and Gilbert (2002a) asked them to produce a model of a process that might be unknown to them (an example was 'how machines that sell soft drinks work') and a model of a scientific idea (an example was 'the occurrence of a chemical reaction'). From their answers, the teachers were then questioned about several specific and related aspects, for example, how they went about the process of modelling; what motivates the building of a model; the possibility of using, and the role for distinct modes of representation; the personal experience, knowledge and attributes of the modeller; the stages followed by a scientists when producing and testing a model; the explanatory and predictive capacities of their models. The analysis of the data showed that such a methodological procedure contributed to the reliability and authenticity of the data since teachers had many opportunities to express ideas related to each aspect (Justi & Gilbert, 2005). As in the case of knowledge about models, these studies indicate that teachers' content knowledge about modelling was far from satisfactory given that they were expected to include this scientific practice in their science teaching. This conclusion has also been assumed, or reached, in studies focused on the development of teachers' knowledge about modelling (Crawford & Cullin, 2004, 2005; Danusso, Testa, & Vicentini, 2010; Davis et al., 2010; Justi, Chamizo, Franco, & Figueirêdo, 2011; Justi & van Driel, 2005c; Kenyon et al., 2011; Schwarz, 2009) that are discussed in a following section.

Contrasting with the limited number of studies into teachers' content knowledge about modelling, there are a larger number of studies involving *PCK about modelling* – which seems to be one of the consequences of researchers' wish to foster the use of MBT in regular classes. Independently of the existence of an explicit focus on 'the construct PCK' as such, the most commonly investigated components of PCK *about modelling* (or *for scientific modelling PCK-SM*, as it is named by some American researchers (Davis et al., 2010; Nelson & Davis, 2012)) have been related to Magnusson et al.'s (1999) definition of PCK for science teaching. In brief, such elements are:

- *orientation towards science teaching using modelling*, which includes knowledge about the role that MBT can play in students' learning, and distinct approaches to MBT, which may result in teachers' intention to teach using the MBT approach (Henze, van Driel, & Verloop, 2007; Justi et al., 2011; Justi & van Driel, 2005c; Nelson & Davis, 2012; Schwarz & Gwekwerere, 2007);
- *knowledge and beliefs about modelling in the science curriculum*, which includes knowing when, how and why to incorporate modelling in science curricula (Henze, van Driel, & Verloop, 2008; Justi et al., 2011; Justi & van Driel, 2005c; Nelson & Davis, 2012; Schwarz, 2009);
- *knowledge and beliefs about instructional strategies that may be used in MBT contexts*. This is a core component of this PCK since it includes knowing when

and how to conduct MBT, when and how to foster the development of students meta-knowledge on models and modelling, and how to engage students in modelling practices (Henze et al., 2008; Justi et al., 2011; Justi & van Driel, 2005b, 2005c; Kenyon et al., 2011; Nelson & Davis, 2012; Schwarz, 2009; Schwarz & Gwekwerere, 2007; Williams & Clement, 2013; Windschitl, Thompson, & Braaten, 2008);
- *knowledge and beliefs about students' understanding of modelling*. This means not only their understanding of modelling practices, epistemology, and nature of models, but also their understanding of the scientific topic(s) involved in a given modelling, and how to supportively discuss the models expressed by students (Henze et al., 2008; Justi et al., 2011; Justi & van Driel, 2005b, 2005c; Kenyon et al., 2011; Nelson & Davis, 2012);
- *knowledge and beliefs about assessment in MBT contexts*. This is about (i) how to assess knowledge and skills acquired by students when participating in MBT contexts, and (ii) how to use modelling activities to assess students relevant knowledge and skills likely to be acquired in those contexts (Henze et al., 2008; Justi et al., 2011; Justi & van Driel, 2005c; Nelson & Davis, 2012).

The studies cited above not only involve more than one aspect of PCK and, sometimes, also aspects of content knowledge. The identification of such aspects is not only undertaken for research purposes. We agree with Kenyon et al. (2011) that the several aspects of teachers' knowledge base "are (or should be) all integrated and closely linked" (p. 4). This claim was corroborated in our fruitless attempt to identify the relationship between teachers' subject knowledge and PCK on modelling in all these studies. We had wanted to classify the view of PCK expressed in each study on teachers' PCK on modelling as being either integrative or transformative (Gess-Newsome, 1999). All the studies to which we had access recognise the existence of PCK as an individual type of knowledge, that is, their authors do nor adopt the integrative view (according to which "PCK does not exist as a domain of knowledge" (Gess-Newsome, 1999, p. 11)). On the other hand, none of them indicates that all other types of knowledge are transformed into PCK such that they would then become the unique teachers' knowledge (a single compound in the chemical reaction system, if we use the analogue domain proposed by Gess-Newsome).

The researchers who have been studying teachers' knowledge about modelling emphasise that other types of knowledge (mainly the pedagogical, the subject content, and the contextual ones) influence, or contribute to, PCK, but without denying the individual existence of any of them. Additionally, some researchers (Figueirêdo & Justi, 2011; Kenyon et al., 2011) explicitly emphasise what we have named a 'dual-transformative interaction' between PCK and other types of knowledge. For them not only do those types of knowledge influence PCK, but also PCK influences them. This is clearly shown that as teachers develop skills and knowledge that support their actions in planning and conducting MBT, they also develop their meta-knowledge about both models and modelling. In our view, this illustrates how complex and challenging is to educate a teacher for MBT. As an attempt to shed some light on such a relevant enterprise, in the following section we discuss how the development of teachers' knowledge on modelling has actually been fostered and supported.

The Development of Teachers' Knowledge About Modelling – Specifics Emerging from the General Guidelines

From the general guidelines for the development of teachers' knowledge previously discussed, we may derive some principles that could be followed in fostering teachers' development about modelling, whether pre-service or in-service. It should be recognised that:

- The learning that is intended to take place should be philosophically and practically congruent with the reasons for including modelling in the science curriculum and with its role in an authentic science education.
- The learning about modelling that takes place should, as far as is possible, occur in social groups composed of peers.
- In order to ensure that the learning is effective, that it relates new ideas to existing ideas and that the conclusions are committed to long-term memory, teachers should have extensive and supported opportunities for reflection on what has been learnt, and on their MBT experiences.
- In order to ensure that generalised knowledge is transformed into effective classroom practice, teachers should have regular access to mentors, that is, colleagues who are both more experienced in MBT and willing to be critically supportive to newcomers.
- In order to support the transition of general ideas into classroom practice, those developing their knowledge and skills in respect of modelling should engage in action research in the area.

Additionally, from the suggestion that teachers' development occurs in three domains (social, personal, and professional) Bell and Gilbert (1996) propose that it can be represented as being marked by three phases during each of which all the types of teachers' knowledge are acquired. In respect of modelling:

- In *phase 1*, a teacher comes of see that being a teacher of modelling is problematic in some respects. It is perceived that address to these problems will involve other teachers, but communicating with them is found difficult. As a way of addressing these uncertainties, a teacher will become prepared to try out new approaches to the teaching of modelling, sensing that traditional practices have been ineffective.

  The place of modelling in science curriculum will have to be appreciated, the nature of effective classroom teaching strategies grasped, and technical competence in formative and summative assessment acquired. In so doing, a teacher will come to clarify their own beliefs about the value of modelling and will learn how the acquisition of its essential components can be supported in the classroom. This will entail a teacher learning about modelling in greater detail than before. This phase places a heavy load of learning and adaptation on the teacher and will probably take place during pre-service education.

- In *phase 2*, a teacher is able to cope with the restraints on the teaching of modelling (for instance, the need to 'cover the syllabus', satisfy external examination criteria) whilst being increasingly willing to try out new approaches. The teacher also develops the ability to undertake mutually professional supportive dialogue with colleagues. During this phase, a coherent and reflective approach to the teaching of modelling is developed.

  A teacher will find ways of teaching modelling in the classroom whilst being aware of competing time pressure. Supporting dialogue with other teachers will have helped how to do this be seen. In so doing, a teacher will acquire a deeper personal understanding of modelling and a wider range of techniques for making material intelligible to students. The overall outcome will be a sense of professional confidence that will be increasingly evident in the classroom. It would be of greatest value if Phase 2 of teacher development in respect of modelling took place in the first years of full-time teaching, although individuals' rate of development will undoubtedly vary depending, amongst other things, on the professional atmosphere in the science department of the school.
- In *phase 3*, a teacher feels sufficiently confident to take the initiative in respect of his/her own professional development. This leads to the active fostering of collaborative ways of working with other teachers and with students. This initiative goes beyond readily available or required opportunities for professional development to establish new structures.

  With Phase 2 successfully completed, a teacher who has come to see the educational significance of modelling in science education will actively seek out new opportunities for using it and/or continue improving their knowledge in the area. This will almost certainly involve teachers in other schools, perhaps also higher-attaining and more science-committed students. Innovative approaches (like the development of animations, and the use of computer-based modelling systems in open-ended project work) and/or more detailed ones (like conducting deeper discussions involving meta-modelling) may also figure in developments.

As shown in the next sub-section, maybe due to the fact that most of the studies in which the development of teachers' knowledge about modelling involve pre-service teachers, they mainly provide examples of teachers development in phases 1 and 2.

The Development of Teachers' Knowledge on Modelling – Some Relevant Studies

Assuming the complexity of teachers' knowledge about models and modelling previously characterised, it does not sound strange that studies that aim at promoting such knowledge development have been focused on some of its specific components. That is the case, for instance, of one of the best-known initial studies that involved specially designed activities in an intervention with pre-service teachers. Crawford and Cullin (2004) moved away from the traditional approach of science methods courses (generally based on studying the pedagogical issues of teaching strategies), and adopted a 'learning through scaffolding' approach. This was done by providing opportunities for pre-service teachers to learn about scientific modelling with the use of the software *Model-It*, that was "designed to support learners in building and testing computer models of dynamic system" (p. 1386). Their main aims were to identify: the pre-service teachers understanding of scientific models and modelling; how such an understanding changed as a result of their participation in the activities; and their intentions concerning teaching about scientific models.

In this study, 14 pre-service science teachers investigated a real-world problem (the relationship between water, soil, and plant growth) before designing and building related computer models that were then tested by running simulations and analysing the results. Data were collected from multiple instruments: pre- and post- open-ended questionnaires (based on Grosslight et al.'s (1991) questions); semi-structured interviews conducted with a sample of the pre-service teachers in order to more deeply explore their written answers; and a series of written questions that stimulated them to reflect on the modelling experience. From the answers to the initial questionnaire, 13 of the participants were classified as level 2 modellers (in the Grosslight et al.'s scale). This result did not change in the final questionnaire, although some more elaborated answers were provided in respect of specific aspects (mainly the use of a scientific model and the language of modelling) and confirmed in the interviews. On the other hand, most of the participants acknowledged the importance of teaching about modelling, but when asked to detail their intentions about doing so, they returned to the comfort of the traditional view of teaching by indicating that modelling activities would be time-consuming and would interfere in the 'coverage of the curriculum'. The authors attributed these slightly positive outcomes to both the fact that, as stated by the participants, they had little, if any, previous experience with scientific modelling, and the short duration of the intervention (less than 12 h distributed over 5 weeks). On account of this, the authors predicted that longer interventions could further enhance the pre-service teachers' knowledge on models and modelling, and that other studies on how to support pre-service teachers to really adopt MBT as part of the teaching approaches in their practices would be necessary. Finally, in terms of such future studies, Crawford and Cullin also emphasised the need to create a comprehensive system of categories for characterising teachers' knowledge about models and modelling, since that of Grosslight et al. could not show the changes in the participants' knowledge that they identified from the contact with them.

Some years later, another study was conducted as part of a science teaching methods course aiming at investigating the development of another component of PCK-SM: teachers' ability to interpret, evaluate, and assess models produced by students. Initially, Nelson and Davis (2012) involved 35 elementary pre-service teachers in modelling-based activities designed to teach evaporation and condensation to elementary students. After some empirical work and the production of their own models, the science educator discussed with them how the models could be analysed and revised using their own evaluations based on three main criteria: consistency with empirical evidence, sense making, and the communicative power of the models. According to the authors, the use of criteria to evaluate models "was revised and reinforced several times in the methods course" (p. 1936). Before participating in the activities, the pre-service teachers answered a pre-test in which they expressed their ideas about models. There is no additional information about this test but, from its answers, four of the pre-service teachers were purposefully selected to be interviewed. During the interviews, they were asked to conduct think-aloud evaluations of the models produced by elementary students, and then to reflect on the criteria they had applied. Data were also gathered from a written assignment that involved the evaluation of a student-generated model, which was done by the pre-service teachers in the middle of the semester.

The results show that the pre-service teachers improved their subject matter knowledge on both areas: 'models and modelling' and 'the scientific topics involved in the activities (evaporation and condensation)' – which the authors view as essential in the evaluation of students' models. They have also improved (i) their model evaluation skills by both focusing their attention on students' ideas as expressed in the model and in applying the discussed evaluation criteria consistently, and (ii) their confidence in their own knowledge and skills in model evaluation. The authors view both aspects as evidence of the development of "the beginnings of PCK" (p. 1953).

Both of these studies focus on single components of PCK about modelling, and were conducted in the context of science methods course, that is, in a situation in which the pre-service teachers were only involved in understanding some aspects of MBT. Therefore, the development of the pre-service teachers in both of them could be said to be focused on the professional domain, and the outcomes classified as falling within phase 1 development.

These studies show that pre-service teachers' actual engagement in modelling is essential to help them if they are to make sense of what they learn in training courses about modelling and MBT. This justifies the authors' recommendation for science educators to invest significant time in science methods course to actually fostering the development of pre-service teachers PCK about modelling. Additionally, it seems crucial that the engagement of pre-service teachers in modelling must not be superficial, that is, that they were not only asked to produce models and evaluate them. As emphasised by Kenyon et al. (2011) when describing efforts to design such kinds of teachers learning experiences (including the one used in Nelson and

Davis (2012) study), pre-service teachers should also be helped to identify what characterise and distinguish MBT from the traditional use of models in science lessons (which only emphasises their explicative and communicative roles). In the successive instructional designs that the American group produced and incorporated into teachers' education programmes in three universities, pre-service teachers were also involved in reflective discussions about: the nature of and purposes for the use of models in science; the construction of a model (including the distinction between conceptual and expressed models); their use, evaluation, and revision; before discussing general issues concerning MBT (including the critical analysis and adaptation of lessons plans to accommodate the modelling perspective). When the evaporation and condensation unit became part of the instructional designs, the pre-service teachers were also asked to use their consensus model to explain other phenomena. In this way, the pre-service teachers were stimulated to teach from the modelling approach, and started developing their PCK about scientific modelling. Finally, in one of the three universities, the pre-service teachers did also have the opportunity to teach parts of the evaporation and condensation unit to their peers, and to reflect on that brief experience. According to Kenyon et al. (2011), such opportunities resulted in improving those pre-service teachers abilities to conduct MBT, in the sophistication of their meta-modelling knowledge, and in increase in the development of their PCK-SM.

In this American teachers' education programme, although the researchers use a nomenclature partially different from the one we have been using, they focused both on the discussions and on the experiences provided to pre-service teachers about modelling using the elements of the four main stages of this process that we identified in one of our initial join papers on teachers' knowledge on modelling (Justi & Gilbert, 2002a), namely: the production of the initial mental model, the expression of that model in any mode of representation, the testing of the model, and its evaluation when used in different contexts.

After completing the study of teachers' knowledge about models and modelling (Justi & Gilbert, 2002a, 2002b, 2003), we identified some aspects as particularly problematic in teachers' knowledge, mainly, the nature of and purposes for the use of models, the use of different models of representation for expressing a model, the use of models for explanatory purposes in teaching (that is, as teaching models), the meaning of the word 'modelling', and its use in science teaching. After that, most of these aspects started to be introduced in the chemistry education methods course in a Brazilian university. There was no data gathering in the context of such initial attempts. However, they were based on the design of a research project that involved serving teachers and which was conducted in The Netherlands. It involved five science teachers who were participating in a 1-year post-graduate teacher education programme, and whose previous professional experienced varied from 3 to 48 months (Justi & van Driel, 2005a, 2005c, 2006). The research questions addressed focused on the fostering and characterising of the development of the teachers' content knowledge, curricular knowledge, and PCK about models and modelling, and on the manifestation of likely changes in the use of teachers' knowledge in their regular classes. It was expected that such discussions would form the basis for the proposal of guideline for teachers' education in this area.

After answering a pre-questionnaire and being interviewed about their initial ideas on models and modelling, the serving teachers took part in 66-h meetings that were held over 2 months. In four of the meetings, they participated in learning activities concerning the aspects mentioned in the previous paragraph, but with two distinctive characteristics:

- the aspects were approached from the teachers' practice perspective (for instance, by taking examples from Dutch textbooks to discuss teaching models); and
- the teachers were involved in a modelling activity as if they were students learning a scientific topic for the first time. This was done by asking them to model a phenomenon that, as they later recognised, they had never thought about before.

The following phases of the project also contributed to differentiate it from other studies in this area at that time: the teachers were asked to select one of the aspects discussed during the meetings on which to base a research project to be conducted in their own regular classes. In order to support their methodological decisions and actions, one of the meetings was focused exclusively on their initial planning. Justi and van Driel were careful with the focus of the discussion so as not to affect the teachers' original research questions, but only to analyse the consistency of, and make suggestions on, their methodological decisions. This specific meeting was followed by a second interview, which aimed at identifying and characterising teachers' knowledge and intentions at that time. As part of the project, the teachers had to collect and analyse relevant data, and to produce a reflective research report whose main points were presented to, and discussed by, the whole group at the last meeting. Finally, after the researchers had read and analysed their reports, the teachers were individually interviewed in order to explain any unclear or incomplete part, to include any additional information about their experience, to be questioned about their then current knowledge about models and modelling, and to evaluate their participation in the whole project. As a whole, the methodological design of this study aimed at contributing to each of the social, personal, and professional development of the serving teachers.

The data from each teacher were analysed by using the Interconnected Model of Teacher Professional Growth (IMTPG), proposed by Clarke and Hollingsworth (2002). This consists of a set of possible non-linear relationships between different domains in which teachers' knowledge may have been revealed: the personal domain (the interaction of teachers initial ideas and beliefs and those acquired during participation in the project); the external domain (the learning activities in which they took part during the meetings); the domain of practice (the teaching situations in which they used or expressed their knowledge during the time they were participating in the project); and the domain of consequence (the outcomes of the domain of everyday classroom practice) – all the relationships being mediated by the processes of reflection and enactment (Justi & van Driel, 2006). Additionally, data from one of the teachers were also used to base a detailed case study of the development of her knowledge (Justi & van Driel, 2005a). The use of the IMTPG produced sufficient and reliable evidence of the complexity and idiosyncrasy of teachers' knowledge and development. The outcomes show if, and how, each part of the instructional design (including the planning, conduction, and reflection on their research projects) contributed to the development of each teacher's content, curricular, and pedagogical content knowledge. Particularly, in the case of their PCK about modelling, the authors focused attention on *the conduction of modelling activities in science teaching* (teachers' planning and actions towards their previous and current experiences, their role, and the discussion of students' generated models), and *the knowledge of students' ideas about modelling* (and their influence on the class discussions). For the three teachers who conducted modelling activities in their research projects, it was possible to show changes in the PCK of two of them, and the emergence of a particularly promising PCK (in the case of one of the teachers who had no previous ideas about MBT (Justi & van Driel, 2005a)). In two of the three cases, the changes resulted from successful experience in terms of students' learning about specific scientific content, as well as in terms of their meta-knowledge about models and modelling. Being consistent with the idiosyncrasies of teachers' development, most of the serving teachers' development reached phase 1, whilst for one of them the development reached phase 2. This was so because her actions when planning and conducting the research project, her reflections expressed in the final report, and her ideas expressed and clearly justified in the final interview, constitute evidence that she had developed a comprehensive content, curricular, and pedagogical content knowledge about models and modelling, being very confident about the use of MBT in specific classroom situations.

From the analysis of the whole process, Justi and van Driel proposed that, when planning and conducting teachers' development programmes in models and modelling, educators should:

- try to characterise teachers' previous content knowledge, curricular knowledge, and PCK about models and modelling in depth so that the subsequent activities and discussions can address issues relevant to those teachers;
- plan activities in which teachers were involved in learning situations analogous to those that students may face in science lessons;
- plan activities that involve elements of their current teaching practice related to models and modelling, and, at the same time, challenge them to analyse their practices;
- justify the new elements of the MBT approach by showing how they are likely to contribute to the enhancement of students' learning;
- request teachers to apply the MBT approach in their lessons, and to investigate the outcomes of such an initiative in terms of students learning;
- provide opportunities for teachers to analyse their new teaching experiences, and reflect on the changes in their knowledge and practices;
- mediate the teachers' reflection processes to some extent at distinct points of time.

Finally, even recognising the importance of the simultaneous consideration of the above-mentioned aspects to make teachers experience a unique learning experience, the authors emphasise one serious limitation of their project: the short time over which teachers' development was expect to occur. According to them, teachers' development in this area should be a long term process since the new experiences on MBT, as well as the teachers' reflections about them, need to be multiple and discussed in depth in order to promote a sustained development of their knowledge that would then support new initiatives. In part, this justifies why, even after the conduct of the research project, that is, during the use of issues related to models and modelling in regular classes, only the development of one teacher reached phase 2.

All recommendations from Justi and van Driel were implemented in a larger subsequent educational programme carried out in Brazil (Figueirêdo & Justi, 2009, 2011). The sample here was composed of nine novice and experienced chemistry teachers (whose professional experience ranged from 1 to 20 years, with a median of 8,5 years) who constituted a collaborative group together with six Master and PhD students in Science Education and one university expert in MBT (RJ). All participants declared their willingness to share their knowledge and teaching experiences, to apply new knowledge in their regular classes, and to critically analyse their own and each other's practices. This means that the programme was developed from a critical-reflective approach whose main premises were:

- the conception of teachers as each of learners, teachers, and researchers in particular moments in the activities;
- the promotion of a real collaboration between teachers and experts, that is, an interaction in which all participants play equally relevant roles through the reciprocal sharing of their knowledge and experiences; and
- a consistent support for teachers during the investigation of, reflections on, and changes into, their practices.

The way these premises were used in the development of the educational programme contributed to stimulating the participants to create and use new teaching activities, to making them feel that their previous knowledge, skills, and experiences were valued, to making them understand that the feedback received was encouraging and could help them (Borko & Putnam, 1996; E.A. Davis, 2003). That level of collaboration is not reached instantaneously. On the contrary, it tends to be reached gradually, and depends directly on the attitudes of all participants. In order to satisfy this specific condition, the teachers' education programme were carried out over one and a half years, during which there were 22 three-h meetings divided in three stages encapsulated in Table 11.1.

Data were gathered with: an initial questionnaire (whose answers have also used as base for some of the activities of the workshops); a second questionnaire completed at the end of stage 1; written materials produced by the teachers during the whole process; video-recording of all meetings and classes involving some of the teachers (those who worked in schools whose head-teachers allowed the recording of the classes to take place); interviews with those teachers at particular moments of the process; and field-notes produced by the researchers after both the meetings and after some of the recorded classes.

Table 11.1 Characterisation of the meetings of the teachers' education programme on MBT

Stage	Number of meetings	General identification	Main characteristics and purposes
1	7	Workshops	A series of activities that aimed at involving teachers in discussing and experiencing issues related to models and modelling, and in analysing modelling-based teaching units previously designed by the research group, as well as some teachers' actions involved in their application in regular classes.
2	5	Design of modelling-based teaching units	Opportunities for the participants, working in small groups, to integrate the knowledge acquired in the previous stage with their previous knowledge and teaching experiences to design modelling-based units for the teaching of chemical concepts selected by them.
3	10	Application of the teaching units and parallel discussion in the group	Collaborative action research (Ponte, 2002) conducted by the teachers in their regular classes, followed by the analysis of the data (videos of classes and materials produced by students); discussions about what happened in all the classes and/or the doubts or problems faced by the teachers when conducting the activities; production of a reflexive report by each teacher; discussions about each teacher's report; and discussions about possible future actions. During the application of the teaching units, there were also sustained phone and e-mail discussions among all participants about specific events and or doubts.

The analysis of such a multiple source of data resulted in a detailed characterisation of the whole process experienced by the teachers. This supported the conclusion that those who actively participated in the entire process developed broad content, curricular, and pedagogical content knowledge about models and modelling from a dynamic and idiosyncratic process of knowledge building, evaluation, reflection, changes, and understanding.

This success was partially attributed to the fact that the learning activities had not been focused simply on acquiring declarative knowledge. On the contrary, they required teachers to use their subject knowledge, curricular, and pedagogical content knowledge in situations directly concerning their teaching practice, and these discussions took place in a friendly and respectful environment. Additionally, as those types of knowledge were being developed, they became crucial in supporting teachers' active participation in the actual design of the modelling-based teaching units, in providing self-confidence in their knowledge and skills in conducting such teaching activities, and in facilitating intense participation in the critical analysis that took place in all the classes. Another factor that contributed to an enhancement of teachers' confidence in MBT were the results they got from their students (in terms of the learning that took place and in their active participation in the lessons). According to one of the teachers, "my students are realising the importance of having opportunities to think, to express and discuss their ideas, and to use them in different contexts". Thus, the evidence obtained by the teachers in their classes, as well as the support they received in the collaborative group, contributed to gradually consolidate all kinds of their knowledge on models and modelling.

In sum, it seems that the simultaneous address to all the guidelines and premises for supporting teachers social, personal, and professional development, as previously discussed in this chapter, was essential for providing a framework that, by being used and continuously discussed and modified throughout the educational programme, constituted a favourable context for the development of all types of knowledge about models and modelling of those teachers.

It is also worth mentioning that:

- during stage 3, one group of teachers produced a paper about the teaching unit they had designed to be presented in a national teachers' conference (Mozzer, Queiroz, & Justi, 2007);
- after the end of the programme, some of the teachers continued to be in touch with the researchers in order to continue to discuss their new experiences on the use of MBT;
- from the participation in this educational programme, three of the teachers decided to enrol themselves on Master courses in Science Education. Two of their subsequent dissertations focused on subjects related to those encountered in the research project: students' and teachers analogical reasoning (Mozzer & Justi, 2012, 2013), and the development of students visualisation skills when participating in MBT activities (Gilbert, Justi, & Queiroz, 2010).

Such outcomes go beyond those regularly obtained in teachers' education programmes. They characterise the development of four of the teachers as eventually reaching phase 3 in the Bell and Gilbert's (1996) model, the development of the other five teachers identified as being in phase 2 by the end of the educational programme.

Looking More Closely at the Classroom Use of Teachers' Knowledge About Modelling

From the experiences of observing, analysing, and discussing the actions of the teachers who participated in that long Brazilian educational programme, as well as the actions of other teachers who conducted MBT when the investigations were focused on students' learning (mainly Maia, 2009; Maia & Justi, 2009; Mendonça & Justi, 2011), it becomes clear that the teachers' role in MBT is an essential one in order to: (i) turn the classes into inquiry environments; (ii) support students in carrying out all the stages of modelling; and (iii) supporting students' development of relevant modelling skills. In order to do so, teachers' content, curricular, and pedagogical content knowledge about models and modelling (as characterised in the earlier section 'Teachers' knowledge on modelling') must be explored in more detail. Such detailed knowledge is needed if it is to be translated into their actions concerning specific aspects of modelling as performed by students (Justi, 2009). The identification of such actions from analyses of regular classes would constitute a reliable basis for planning teachers' development programmes concerning MBT. Therefore, from a comparative analysis of data available until 2009, Justi (2009, 2013) tried to identify core actions that should be performed by teachers when conducting MBT.

A similar rationale was adopted for the a recent work of Williams and Clement (2013, 2014, 2015), in which they tried, firstly, to identify the cognitive strategies used by experienced teachers in model-based science classes in order to secondly, later, support the development of pre-service teachers' skills in teaching from such an approach. According to the authors, such strategies involve "questions and comments that respond to specific strengths and weaknesses in the ideas being expressed by students" (Williams & Clement, 2015, p. 84). From the analysis of data generated in two teachers' classes, they propose the existence of three levels of teaching strategies: one of *dialogical strategies*, and two levels of *cognitive strategies*, respectively macro and micro. The dialogical strategies relate to support the communication of students ideas in class discussions (for instance, summarising students ideas, "redirecting questions back to students rather than providing answers, focusing attention on conflicts and differences of opinion" (Williams & Clement, 2013, p. 5)). On the other hand, the *macro cognitive strategies* include all teachers' statements that help students express their ideas concerning Observations (O) of evidence and/or prior knowledge that seems to be the base for model construction; Generation (G) of a model; Evaluation (E) of a model under discussion; and Modification (M) of such a model. The *micro cognitive strategies* are sub-strategies for achieving one of the broader macro OGEM strategies. Thirty-nine micro cognitive strategies were initially identified, and then amalgamated into 15 key micro strategies.

Table 11.2 provides a comparison between the teachers' core actions as identified by Justi and the micro strategies as identified by Williams and Clement. In this table, there is a correspondence between the 'modelling stages' (that are expressed using the nomenclature adopted before the publication of this book, that is, in accordance with Fig. 2.2) and 'macro strategies'.

As emphasised in Chaps. 2 and 4, the two MBT proposals that basis the work of the above authors have some particularities. In the current context, the most important ones are: the different stages at which modelling is assumed to be carried out; and the fact that the activities that comprise the teaching units designed from the Model of Modelling approach support, or complement, the teachers' actions (for instance, by providing empirical experiments for the acquisition of information that may be used to produce or test a given model, or by asking for the generation of analogies at specific times). By ignoring such differences, Table 11.2 shows that the teacher's roles in both proposals are quite similar. In some sense, this would be expected since both of them are grounded in the same constructivist philosophical principals. However, the major conclusion we draw from the analysis of Table 11.2 it that the convergence of teachers' actions shown indicates the need to broaden the current accepted view of PCK on modelling. In this sense, for instance, the element 'knowledge and beliefs about instructional strategies that may be used in a science teaching context' could not be viewed from the general phrase 'knowledge on how to conduct MBT' (that is one of the current accepted elements of PCK on modelling). But it could rather be understood from one element that is at the core of PCK on modelling: the meaning of conducting MBT. Such a more detailed view of PCK about modelling could also support teachers' educators in:

- the identification of key elements involved in the social, personal, and professional development of teachers, since some constraints on these processes may be related to specific actions that teachers are expected to perform, or to the consequences of such actions in the classroom context;
- the design of learning activities that may support the evolving of teachers' knowledge from phases 1–3.

An attempt to design a teachers' educational programme from this approach was made by Williams and Clement (2013, 2014). They have used videos from classes

Table 11.2 Comparison between Justi's key teachers' actions in MBT and Williams and Clement's modelling teaching strategies

Justi's key teachers' actions in MBT		Williams and Clement's modelling teachers' strategies	
Modelling stages	Teachers' key actions	Macro strategies	Micro strategies
Production	To support students' engagement in thinking about a given phenomenon or system (by explaining of giving a counter-example of some aspect of the phenomenon or system).	Observation	Requests or provides observations.
			Requests or provides diagram to help students recall results of an experiment.
		Generation	Requests or provides the initiation of model construction.
	To help students recall their previous knowledge or models and establish relationships between them and the system under study (by questioning students about, and to compare, them).		Requests or provides a model element to explain specific observation.
			Requests or provides new detail or elaboration of the model.
			Requests or provides spatial direction of effect.
			Requests or provides an analogy.
Expression	To support students' clear communication of their models (by asking questions about: the ideas expressed in their models, the modes of representation used by them, and the codes of representation used for each mode).		

(continued)

Table 11.2 (continued)

Justi's key teachers' actions in MBT		Williams and Clement's modelling teachers' strategies	
Test	To support the development of students' abilities of conducting thought and empirical experiments.	Evaluation	Requests or provides evidence to support or refute a model.
			Requests or provides the design of an experiment or thought experiment.
	To help students in analysing the results of both thought and empirical experiments (by questioning students about their understanding of the results).		Requests or provides running a model for prediction or evaluation.
			Requests or provides a discrepant question or discrepant event.
		Modification	Requests or provides additions or changes to the model.
			Requests or provides differentiation between elements of models.
			Requests or provides integration of two models or concepts.
	To help students realise possible incoherence between their models and current evidence or previous knowledge (by questioning students about possible interpretations for the results of the tests).		Requests or provides repair to or refinement of the language describing the model.
Evaluation	To foster students to elaborate questions that could result in testing the model in different contexts.		

After Justi (2009, 2013); Williams and Clement (2015)

conducted by an experienced teacher, discussed some diagrammatic representations produced from the interpretation of such classes from the three levels of teaching strategies, and provided opportunities for pre-service teachers to try out some of the strategies through peer-to-peer micro teaching sessions on a secondary class. Such mini-lessons were video-recorded and then analysed and discussed by the whole group. After this 8-week instructional unit, a comparison between a pre- and a post-instruction survey showed a significant change in the pre-service teachers' views about model-based whole class discussions. As emphasised by the authors, this is only one step towards the complex process of educating teachers.

Other authors (for instance, Windschitl et al., 2008) have identified distinct, but equally important, elements that should be focused on attempts to promote the development of teachers' PCK in this area. Although it is not possible to identify changes in the knowledge of the pre-service teachers who participated in

Williams and Clements study in terms of different phases, it seems that those pre-service teachers have develop a strong phase 1 knowledge, which could base promising changes.

Concluding Remarks

This chapter shows that becoming a teacher who effectively facilitates students' learning in MBT context involves the development of a series of distinct and integrated knowledge and skills. This also requires a shift in teachers' meta-knowledge about models and modelling since modelling has to be viewed not only from a communicative perspective. Rather, teacher must understand modelling as a cognitive process that mainly aims at generating knowledge in both, science and science teaching.

It is clear that this is an on-going enterprise that, having started in science methods course, may continue for teachers' professional life. This would be a consequence of the permanent knowledge growth of both areas: MBT and teachers' actions. As it was emphasised by a Mexican teacher who participated in a professional development project similar to the one conducted in Brazil (characterised in Table 11.1):

> We know this work does not finish here. This is only the beginning, and there are a lot of missing issues to be learnt before we can conduct MBT by taking all relevant details into account. This project made us think, reflect, explore our possibilities, recognise our limitations and explore previously unknown limits of our capabilities. As teachers, now it is our duty and responsibility to continue searching for new ways to improve our teaching performance in MBT contents. (Justi et al., 2011, p. 423)

Therefore, those involved in the challenge of enhancing teachers' education should view their enterprise as an evolutionary and non-predictive one. As in any modelling case, a given test, a different context, or a creative way to analyse them may mean significant changes in approaches and/or directions. Some ideas to encourage researchers and teachers educators are discussed in the next chapter.

References

Abell, S. K. (2008). Twenty years later: Does pedagogical content knowledge remain a useful idea? *International Journal of Science Education, 30*(10), 1405–1416.

Barnett, J., & Hodson, D. (2001). Pedagogical content knowledge: Toward a fuller understanding of what good science teachers know. *Science Education, 85*(4), 426–453.

Bell, B., & Gilbert, J. K. (1996). *Teacher development: A model from science education*. London, UK: Falmer.

Berry, A., Loughran, J., & van Driel, J. (2008). Revisiting the roots of pedagogical content knowledge. *International Journal of Science Education, 30*(10), 1271–1279.

Borko, H. (2004). Professional development and teacher learning: Mapping the terrain. *Educational Researcher, 33*(8), 3–15.

Borko, H., & Putnam, R. (1996). Learning to teach. In D. C. Berliner & R. C. Calfee (Eds.), *Handbook of educational psychology* (pp. 673–708). New York, NY: Macmillan.

Burbank, M. D., & Kauchak, D. (2003). An alternative model for professional development: Investigations into effective collaboration. *Teaching and Teacher Education, 19*(5), 499–514.

Calderhead, J., & Gates, P. (1993). Introduction. In J. Calderhead & P. Gates (Eds.), *Conceptualizing reflection in teacher development* (pp. 1–10). London, UK: The Falmer Press.

Capps, D., & Crawford, B. A. (2013). Inquiry-based professional development: What does it take to support teachers in learning about inquiry and nature of science? *International Journal of Science Education, 35*(12), 1947–1978.

Clarke, D., & Hollingsworth, H. (2002). Elaborating a model of teacher professional growth. *Teaching and Teacher Education, 18*, 947–967.

Crawford, B. A., & Cullin, M. J. (2004). Supporting prospective teachers' conceptions of modeling in science. *International Journal of Science Education, 26*(11), 1379–1401.

Crawford, B. A., & Cullin, M. J. (2005). Dynamic assessments of preservice teachers' knowledge of models and modelling. In K. Boersma, M. Goedhart, O. de Jong, & H. Eijkelhof (Eds.), *Research and quality of science education* (pp. 309–323). Dordrecht, The Netherlands: Springer.

Danusso, L., Testa, I., & Vicentini, M. (2010). Improving prospective teachers' knowledge about scientific models and modelling: Design and evaluation of a teacher educational intervention. *International Journal of Science Education, 32*(7), 871–905.

Davis, K. S. (2003). "Change is hard": What science teachers are telling us about reform and teacher learning of innovative practices. *Science Education, 87*(1), 3–30.

Davis, E. A. (2003). Knowledge integration in science teaching: Analysing teachers' knowledge development. *Research in Science Education, 34*(1), 21–53.

Davis, E. A., Kenyon, L., Hug, B., Nelson, M. M., Beyer, C., Schwarz, C. V., & Reiser, B. J. (2008). *MoDeLS: Designing supports for teachers using scientific modeling*. Paper presented at the Association for Science Teacher Education, St. Louis, MO.

Davis, E. A., Nelson, E., Hug, B., Kenyon, L., Cotterman, M., & Teo, T. W. (2010). *Preservice teachers and scientific modeling: Synthesizing results of a multi-year multi-site project*. Paper presented at the ASTE international meeting, Sacramento, CA.

Figueirêdo, K. L., & Justi, R. (2009). Desenvolvimento do PCK de professores sobre modelagem a partir da realização de pesquisa-ação em um grupo colaborativo [Development of teachers' PCK on modelling from the conduction of action research in the context of a collaborative group]. *Enseñanza de las Ciencias, 27*(extra), 702–706.

Figueirêdo, K. L., & Justi, R. (2011). Uma Proposta de Formação Continuada de Professores de Ciências Buscando Inovação, Autonomia e Colaboração a partir de Referenciais Integrados [A proposal for science teachers education aiming at innovation, autonomy, and collaboration]. *Revista Brasileira de Pesquisa em Educação em Ciências [Brazilian Journal of Research in Science Education], 11*(1), 169–190.

Friedrichsen, P., van Driel, J., & Abell, S. K. (2010). Taking a closer look at science teaching orientations. *Science Education, 95*(2), 358–376.

Gess-Newsome, J. (1999). Pedagogical content knowledge: An introduction and orientation. In J. Gess-Newsome & N. G. Lederman (Eds.), *Examining pedagogical content knowledge: The construct and its implications for science education* (pp. 3–17). Dordrecht, The Netherlands: Kluwer.

Gess-Newsome, J. (2013). *The PCK Summit consensus model and definition of pedagogical content knowledge*. Paper presented at the X conference of the European Science Education Research Association, Nicosia, Cyprus.

Gess-Newsome, J., & Lederman, N. G. (1999). *Examining pedagogical content knowledge: The constructs and its implications for science education*. Dordrecht, The Netherlands: Kluwer.

Gilbert, J. K., Justi, R., & Queiroz, A. S. (2010). The use of a Model of Modelling to develop visualization during the learning of ionic bonding. In M. F. Taşar & G. Çakmakcı (Eds.), *Contemporary science education research: International perspectives* (pp. 43–51). Ankara, Turkey: Pegem Akademi.

Grosslight, L., Unger, C., Jay, E., & Smith, C. L. (1991). Understanding models and their use in science: Conceptions of middle and high school students and experts. *Journal of Research in Science Teaching, 28*(9), 799–822.

Grossman, P. L. (1990). *The making of a teacher: Teacher knowledge and teacher education.* New York, NY: Teachers College Press.

Henze, I., van Driel, J., & Verloop, N. (2007). The change of science teachers' personal knowledge about teaching models and modelling in the context of science education reform. *International Journal of Science Education, 29*(15), 1819–1846.

Henze, I., van Driel, J., & Verloop, N. (2008). Development of experiences science teachers' pedagogical content knowledge of models of the solar system and the universe. *International Journal of Science Education, 30*(10), 1321–1342.

Justi, R. (2009). Learning how to model in science classroom: Key teacher's role in supporting the development of students' modelling skills. *Educacion Quimica, 20*(1), 32–40.

Justi, R. (2013). *Co-construction of knowledge in a modeling-based teaching context.* Paper presented at the NARST 2013 annual international conference, Rio Grande, Puerto Rico.

Justi, R., Chamizo, J. A., Franco, A. G., & Figueirêdo, K. L. (2011). Experiencias de formación de profesores latinoamericanos de ciencias sobre modelos y modelaje [Experiences on Latin-American science teachers' education on models and modelling]. *Enseñanza de las Ciencias, 29*(3), 413–426.

Justi, R., & Gilbert, J. K. (2002a). Modelling, teachers' views on the nature of modelling, implications for the education of modellers. *International Journal of Science Education, 24*(4), 369–387.

Justi, R., & Gilbert, J. K. (2002b). Science teachers' knowledge about and attitudes towards the use of models and modelling in learning science. *International Journal of Science Education, 24*(12), 1273–1292.

Justi, R., & Gilbert, J. K. (2003). Teachers' views on the nature of models. *International Journal of Science Education, 25*(11), 1369–1386.

Justi, R., & Gilbert, J. K. (2005). Investigating teachers' ideas about models and modelling – some issues of authenticity. In K. Boersma, M. Goedhart, O. de Jong, & H. Eijkelhof (Eds.), *Research and the quality of science education* (pp. 325–335). Drodrecht, The Netherlands: Springer.

Justi, R., & van Driel, J. (2005a). A case study of the development of a beginning chemistry teacher's knowledge about models and modelling. *Research in Science Education, 35*(2&3), 197–219.

Justi, R., & van Driel, J. (2005b). Developing science teachers' knowledge on models and modelling. In D. Beijaard, P. C. Meijer, G. Morine-Dershimer, & H. Tillema (Eds.), *Teacher professional development in changing conditions* (pp. 165–180). Dordrecht, The Netherlands: Springer.

Justi, R., & van Driel, J. (2005c). The development of science teachers' knowledge on models and modelling: Promoting, characterizing and understanding the process. *International Journal of Science Education, 27*(5), 549–573.

Justi, R., & van Driel, J. (2006). The use of the interconnected model of teacher professional growth for understanding the development of science teachers' knowledge on models and modelling. *Teaching and Teacher Education, 22*(4), 437–450.

Kenyon, L., Davis, E. A., & Hug, B. (2011). Design approaches to support preservice teachers in scientific modeling. *Journal of Science Teacher Education, 22*(1), 1–21.

Kind, V. (2009). Pedagogical content knowledge in science education: Perspectives and potential for progress. *Studies in Science Education, 45*(2), 169–204.

Kloser, M. (2014). Identifying a core set of science teaching practices: A Delphi expert panel approach. *Journal of Research in Science Teaching, 51*(9), 1185–1217.

Krell, M., & Krüger, D. (2015). Testing models: A key aspect to promote teaching activities related to models and modelling in biology lessons? *Journal of Biological Education.* doi:10.1080/00219266.2015.1028570.

Lin, H.-S., Hong, Z.-R., Yang, K.-K., & Lee, S.-T. (2013). The impact of collaborative reflections on teachers' inquiry teaching. *International Journal of Science Education, 35*(18), 3095–3116.

Loughran, J. (2006). *Developing a pedagogy of teacher education: Understanding teaching and learning about teaching*. London, UK: Routledge.

Loughran, J., Berry, A., & Mulhall, P. (Eds.). (2006). *Understanding and developing science teachers' pedagogical content knowledge*. Rotterdam, The Netherlands: Sense.

Loughran, J., Mulhall, P., & Berry, A. (2004). In search of pedagogical content knowledge in science: Developing ways of articulating and documenting professional practice. *Journal of Research in Science Teaching, 41*(4), 370–391.

Magnusson, S., Krajcik, J., & Borko, H. (1999). Nature, sources and development of pedagogical content knowledge for science teaching. In J. Gess-Newsome & N. G. Lederman (Eds.), *Examining pedagogical content knowledge: The construct and its implications for science education* (pp. 95–132). Dordrecht, The Netherlands: Kluwer.

Maia, P. F. (2009). *Habilidades Investigativas no Ensino Fundamentado em Modelagem [Investigative skills in modelling-based teaching]*. PhD thesis, Universidade Federal de Minas Gerais, Belo Horizonte, Brazil.

Maia, P. F., & Justi, R. (2009). Learning of chemical equilibrium through modelling-based teaching. *International Journal of Science Education, 31*(5), 603–630.

Mendonça, P. C. C., & Justi, R. (2011). Contributions of the Model of Modelling diagram to the learning of ionic bonding: Analysis of a case study. *Research in Science Education, 41*(4), 479–503.

Mozzer, N. B., & Justi, R. (2012). Students' pre- and post-teaching analogical reasoning when they draw their analogies. *International Journal of Science Education, 34*(3), 429–458.

Mozzer, N. B., & Justi, R. (2013). Science teachers' analogical reasoning. *Research in Science Education, 43*(4), 1689–1713.

Mozzer, N. B., Queiroz, A. S., & Justi, R. (2007). *Proposta de Ensino de Introdução ao Tema Interações Intermoleculares via Modelagem [A proposal of modelling-based activities for the teaching of intermolecular interactions]*. Paper presented at the VI Encontro Nacional de Pesquisa em Educação em Ciências [VI Brazilian conference on reearch in science education], Florianópolis.

Nelson, M. M., & Davis, E. A. (2012). Preservice elementary teachers' evaluations of elementary students' scientific models: An aspect of pedagogical content knowledge for scientific modeling. *International Journal of Science Education, 34*(12), 1931–1959.

Nilsson, P. (2008). Teaching for understanding: The complex nature of pedagogical content knowledge in pre-service education. *International Journal of Science Education, 30*(10), 1281–1299.

Park, S., & Oliver, J. S. (2008). Revisiting the conceptualisation of pedagogical content knowledge (PCK): PCK as a conceptual tool to understand teachers as professionals. *Research in Science Education, 38*(3), 261–284.

Ponte, P. (2002). How teachers become action researchers and how teacher educators become their facilitators. *Educational Action Research, 10*, 399–422.

Putnam, R. T., & Borko, H. (2000). What do new views of knowledge and thinking have to say about research on teacher learning? *Educational Researcher, 29*(1), 4–15.

Reiser, B. J. (2013). *What professional development strategies are needed for successful implementation of the next generation science standards?* Paper presented at the invitational research symposium on science assessment, Washington, D.C.

Schön, D. A. (1987). *Educating the reflective practitioner – Toward a new design for teaching and learning in the professions*. San Francisco, CA: Wiley.

Schön, D. A. (1991). *The reflective practitioner – How professionals think in action* (2nd ed.). London, UK: Ashgate.

Schwarz, C. V. (2009). Developing preservice elementary teachers' knowledge and practices through modeling-centered scientific inquiry. *Science Education, 93*(4), 720–744.

Schwarz, C. V., & Gwekwerere, Y. N. (2007). Using a guided inquiry and modeling instructional framework (EIMA) to support preservice K-8 science teaching. *Science Education, 91*(1), 158–186.

Shulman, L. S. (1986). Those who understand: Knowledge growth in teaching. *Educational Researcher, 15*, 4–14.

Shulman, L. S. (1987). Knowledge and teaching: Foundations of the new reform. *Harvard Educational Review, 51*(1), 1–22.

Sperandeo-Mineo, R. M., Fazio, C., & Tarantino, G. (2005). Pedagogical content knowledge development and pre-service physics teacher education: A case study. *Research in Science Education, 36*(3), 235–264.

van Driel, J., & Berry, A. (2012). Teacher professional development focusing on pedagogical content knowledge. *Educational Researcher, 41*(1), 26–28.

van Driel, J., Meirink, J. A., van Veen, K., & Zwart, R. C. (2012). Current trends and missing links in studies on teacher professional development in science education: A review of design features and quality of research. *Studies in Science Education, 48*(2), 129–160.

van Driel, J., & Verloop, N. (1999). Teachers' knowledge of model and modelling in science. *International Journal of Science Education, 21*(11), 1141–1154.

van Driel, J., & Verloop, N. (2002). Experienced teachers' knowledge of teaching and learning of models and modelling in science education. *International Journal of Science Education, 24*(12), 1255–1272.

Williams, E. G., & Clement, J. J. (2013). *From research to practice: Fostering pre-service science teachers' skills in facilitating effective whole class discussions*. Paper presented at the NARST 2013 annual international conference, Rio Grande, Puerto Rico.

Williams, E. G., & Clement, J. J. (2014). *Using research on cognitive discussion strategies to support pre-service science teachers' model-based teaching skills*. Paper presented at the National Association for Research in Science Teaching Conference, Pittsburgh, PA.

Williams, E. G., & Clement, J. J. (2015). Identifying multiple levels of discussion-based teaching strategies for constructing scientific models. *International Journal of Science Education, 37*(1), 82–107.

Windschitl, M., & Thompson, J. (2006). Transcending simple forms of school science investigation: The impact of preservice instruction on teachers' understanding of model-based inquiry. *American Educational Research Journal, 43*(4), 783–835.

Windschitl, M., Thompson, J., & Braaten, M. (2008). How novice science teachers appropriate epistemic discourses around model-based inquiry for use in classrooms. *Cognition and Instruction, 26*(3), 310–378.

Zembal-Saul, C., Blumenfeld, P., & Krajcik, J. (2000). Influence of guided cycles of planning, teaching, and reflection on prospective elementary teachers' science content representations. *Journal of Research in Science Teaching, 37*(4), 318–339.

Chapter 12
Modelling-Based Teaching and Learning: Current Challenges and Novel Perspectives

Abstract In this chapter we draw together the themes, identified in earlier chapters, which must be addressed if 'modelling-based teaching' is become part of the professional repertoire of all school science teachers. We also tentatively identify the professional agencies that may be best placed, in terms of their expertise, to do so. Finally, we summarise the novel perspectives that we have advanced about these and suggest how these may be helpful in addressing the challenges identified.

The Challenges and the Challengers

In order to structure this discussion, we remind readers that the successive themes addressed in this book were:

- Chapter 1: the challenges that science education currently faces, together with the assertion that an education in and about modelling can help meet these challenges;
- Chapter 2: the notion of 'model' and the knowledge and skills that contribute to the production and validation of models;
- Chapter 3: the notion of 'authentic learning in science' together with an evaluation of how modelling can contribute to that authenticity;
- Chapter 4: an exploration of the meaning of MBT together with the presentation of a 'model of modelling';
- Chapter 5: an exploration of the scope and limitations of the meaning of the words 'concept' and 'model' as these are often confused in the literature;
- Chapter 6: the use of argumentation in the acts of creating and validating models;
- Chapter 7: the contribution that visualisation makes to the creation of models;
- Chapter 8: the central role of analogies in modelling-based teaching;
- Chapter 9: the way that modelling contributes to the core curricular aim of 'understanding the scientific enterprise';

J.K. Gilbert, R. Justi, *Modelling-based Teaching in Science Education*, Models and Modeling in Science Education 9, DOI 10.1007/978-3-319-29039-3_12

- Chapter 10: the structure of a 'learning progression' for modelling;
- Chapter 11: the professional development of teachers needed to implement modelling-based teaching.

As we explored those themes, it became increasingly apparent that some major aspects of them had not been subject to the research and development that would make the universal introduction of MBT a realistic proposition. In revisiting those aspects and tentatively suggesting some of what is to be done, we are aware that the various agencies that are involved in the massive enterprise that is science education each have distinctive, but often overlapping, expertise to contribute to that research and development. We have therefore structured our analysis within those agencies, which we see to be seven in number.

Curriculum Designers

The main task of this group is to signal the importance of MBT to all the other professionally agencies active in the field of science education and to summarise all the main ideas that must be included in their treatment of it.

We must assume that the general principles of what might be termed 'constructivist teaching' are widely known. The issue is then how to adapt those principles to the case of MBT. In particular, curriculum designers would need to set out:

- the broad principles that should guide the design of MBT activities intended to support students' simultaneous involvement in the full range of epistemic practices that comprise modelling (Chap. 4);
- the broad principles that should guide the design of activities to support students' development of argumentation (Chap. 6), visualisation (Chap. 7), and analogical reasoning (Chap. 8);
- the way that the essential attributes of such activities contribute to students' (i) learning of scientific or socio-scientific issues, (ii) development of a comprehensive view about science, and (iii) development of modelling competence (Chap. 4);
- the ways that 'artefactual' view enables the meaning of 'model' and 'concept' to converge (Chap. 5);
- the variants in these principles that are needed to accommodate the differences between the educational environments of primary, secondary, and tertiary, science classrooms (Chap. 10).

Science Education Researchers

This group collectively has the expertise necessary to use the ideas of education, philosophy, psychology, and sociology, to identify and explore in depth innovative questions involving (i) the elements of MBT and the relationships between them, and (ii) creative ways to educate teachers who may be able to successfully conduct MBT activities.

Some important issues that can best be tackled by science education researchers are:

- the extent to which an increased focus on models and modelling addresses the disengagement of students from the sciences (Chap. 1);
- the contribution of a knowledge about modelling to the 'scientific literacy' of all students (Chap. 1);
- the value of the 'Model of Modelling v2' as a basis for MBT (Chap. 4);
- the contribution of MBT to the realisation of situated cognition in science education (Chap. 4);
- the skills of argumentation that are needed by students if they are to be able to engage in modelling activities in contexts of their own choosing after the conclusion of formal MBT (Chap. 6);
- the relationships between the stages of modelling and the visualisation skills and abilities needed to accomplish them (Chap. 7);
- the relationships between LPs concerning the distinct practices and cognitive processes involved in MBT (Chap. 10).

Advanced Students of Science Education and Curriculum Design

This group, almost certainly composed of people with some experience as classroom science teachers, will be able to focus in depth on particular questions identified by the science education researchers.

This group will be best placed to engage in that detailed empirical work concerning:

- the contribution of a knowledge about models and modelling to the cultural capital of students (Chap. 1);
- the ways that the explicit teaching of argumentation can be related to activities within the distinctive stages of modelling (Chap. 6);
- the relationships that can be established between students' epistemological views about science and the range and quality of their argumentative skills (Chap. 6);
- the relationships between such a development and the use of specific modes of representation of models (Chap. 7);
- the ways in which the use of computer-based modelling activities are similar and different in their impact of student learning from the use of other types of MBT activities (Chap. 7);
- the relationships that can be established between students' epistemological views about science and their performance in modelling activities (Chap. 9);

- the ways that the gradual development of meta-visual competence is fostered by specific aspects of MBT (Chap. 10);
- the degree to which authentic MBT can be attained in science classrooms (Chap. 10).

Teacher Educators

In order to effectively contribute to the development of pre- and/or in-service teachers' relevant knowledge on MTB, the teacher educators will be individuals who have (i) extensive experience as practitioners of MBT, (ii) a grasp of how to both introduce those ideas to pre-service and in-service science teachers, (iii) the ability to convince them that MBT is educationally worthwhile.

Important issues for this group are:

- the approaches to MBT needed to ensure that engagement in professional development activities will lead to teachers being both willing and able to engage in MBT, as opposed to merely knowing about it (Chap. 11);
- the phasing of education in respect of MBT across the professional development of teachers that will ensure the maximum impact on their classroom practice (Chap. 11);
- the focus of teachers' education on a comprehensive view of modelling that includes all the other processes involved on it (Chaps. 6, 7, 8) and that contribute to the development of a more consistent view about science (Chap. 9). In other words, teachers' educators have to contribute to pre-service teachers understanding about the distinct LPs characterised in Chap. 10, so that they could be able to address them in their teaching practice, thus supporting the occurrence of simultaneous LPs (as represented in Fig. 10.1).

Practicing Classroom Teachers

The group, consisting of primary (elementary), secondary (high school), and tertiary (university) teachers would have the key role of ensuring that all the other expert cadres 'keep their feet on the ground', that is, ensuring that what is proposed, or indeed, mandated, is realistic for the great majority of classrooms.

The big issues here are:

- the similarities and differences in MBT as practiced with students from different age cohorts in regular classes (Chap. 10);
- the progression of that practice across the increasing age cohorts (Chap. 10);

- the characteristics of authentic real world problems and contexts that may be addressed in MBT as a way of supporting students' learning of issues relevant to their education, both general and scientific, as twenty-first century citizens (Chap. 10);
- the creation of valid and reliable instruments for the formative assessment of students' learning during MBT in regular classes (Chap. 10).

Public Examiners

Given that all educational system are subject to the rigours of 'accountability' and that this is manifest in student assessment, the cadre of public examiners has great importance, for what is included about modelling in examination syllabi is what will be taught in classrooms.

The key question, so far virtually unaddressed, is the ways that the knowledge and skills of MBT can be validly, reliably, and economically, assessed for both formative and summative purposes (Chap. 10).

Textbook Designers

In many national educational systems, science teaching consists of following the development of ideas in the order and manner laid out in approved textbooks. If MBT is to be widely practiced, what it entails and how it can be practiced must appear prominently in such textbooks.

Questions of importance are:

- the extent that MBT is actively supported by textbooks at the moment (Chap. 1);
- the relation of the treatment of MBT to that on other material in textbooks (Chap. 1);
- the topics for which MBT would be particularly relevant in supporting students' learning (Chap. 3);
- the ways to simultaneously emphasise students' learning of scientific content, and the development of their: meta-knowledge on models and modelling (Chap. 5), broader understanding of science (Chap. 9), performance of the several practices and practices involved in modelling (Chap. 10).

Assuming that these issues can be addressed to a satisfactory degree (a large assumption!), approaches to meeting challenges to teachers and teacher educators (arguably the most important groups of professionals) can be identified as follows.

Re-Dimensioning the Challenge of Educating Students from a MBT Perspective

In Chap. 10, following an analysis of the current literature on students' learning of modelling, we predicted that a potentially successful generic learning progression on modelling must:

- address both models and modelling;
- be based on an address to phenomena that students see (or at least passively accept) as being of authentic interest;
- focus on the construction and evaluation of models by the students themselves;
- involve the provision of scaffolding for student activity by the teacher;
- involve students in reflecting on what they have done during the process of modelling.

After taking into account all the issues discussed in Chaps. 6, 7, 8, and 9, we concluded that the attainment of an LP in models and modelling will have to be intertwined with the attainment of an LP for each of argumentation, visualisation, analogical reasoning, and learning about science. It therefore seems that the main challenges to the education of students from the modelling-based perspective, as characterised in this book, must involve the design and assessment of activities within such an LP. The curricular integration that such an overall LP infers will require the design of a set of coherent activities. This requirement means much more than simultaneously taking into account the aspects that comprise the competence in each of the constituent LPs as identified in consecutive sections of Chap. 10. Whilst the separate components of the LPs have been identified earlier in this book, their integration into an overall curriculum still remains to be done. As such, it will be the focus of a substantial future programme of curricular development.

In a sub-section of Chap. 10 ('Model 1: An explicit and progressive exposure to competence in modelling'), we tried to discuss the explicit provision of a progressive exposure to all the components of a capability in models and modelling for each of the five modelling approaches. For the four simpler modelling approaches, this was done in the light of the current literature. The last and more complex approach in that sequence – learning to construct a model de novo – was discussed from the proposition of a sketch to a whole progression concerning learning to construct a model de novo (Fig. 10.1). However, in order to change such a sketch into a real proposal on how to support – and explain – students' modelling-based learning, we view a substantial and sustained programme of research and development funding as absolutely needed. Then, a much deeper knowledge on all the five approaches to modelling would be increased, and a curriculum based on such approaches might be available.

Re-Dimensioning the Challenge of Educating Teachers to Facilitate MBT

Our analysis of the literature on teachers' knowledge and development about models and modelling (Chap. 11) was conducted from a strongly critical perspective after we had written the other chapters of this book. We conclude that what has been publishing in the area is just the tip of the iceberg of what is needed. We put forward two justifications for this assertion.

First, in brief, in this book we assume that modelling can best be seen as being the production of artefacts that can be used in the many facts of scientific practices and in distinct ways. From what we discussed in Chap. 2 about the nature of modelling, and from the design the 'Model of Modelling v2' (Figs. 2.3, 2.4, and 2.5), it emerged that modelling is a cyclic, non-linear, and non-predetermined process of creating a proto-model, expressing the proto-model in any mode of representation, empirically and/or thoughtfully testing the model (and modifying it, when necessary), and evaluating it in order to identify its scope and limitations. This view supported the identification of a series of skills and abilities that one needs in order to perform each of the stages of modelling (Table 4.1). Additionally, some key epistemic practices were associated with the performance of this process: the use of analogical reasoning, the use of imagistic representations, the design and run of thought experiments, and argumentation. From the focus on such practices in other Chaps. (mainly 6, 7, and 8), we identified several other domains of knowledge, skills and abilities that are necessary for a genuine experience of modelling that could contribute to an individual's understanding of the epistemic foundations of science, as well as of the cognitive and social dimensions of science (as discussed in Chap. 9).

Therefore, by assuming that, in order to plan and conduct MBT, teachers must have a comprehensive understanding of models and modelling, it emerges that science teachers (in any stages of their professional career) have to build a flexible and dynamic network of knowledge, skills and abilities related to all the elements and epistemic practices involved in modelling. This means much more than what the literature describes as teachers having to learn about models and modelling in exposition-based 'science methods' courses. That is certainly necessary but not sufficient. Teachers must also develop *a comprehensive* understanding of MBT itself:

- its role in supporting students' authentic science learning (as characterised in Chap. 4);
- when, how, and why to include MBT in science curricula at distinct school levels;
- how and why students understand (or do not understand) each of the elements related to modelling;
- how and why students are able to perform each of the related skills and abilities;
- how and why the participation in MBT contributes to students learning about science.

Teachers must also develop in-depth knowledge and considerable skills and abilities concerning every single action necessary:

- to involve students in MBT activities (a sample of which are presented in Chap. 11);
- to support students' performance of each of the skills and abilities involved in specific modelling stages;
- to deal with students' assessment issues in MBT contexts; and
- to make MBT a powerful instrument for supporting an extensive, authentic, and functional science learning.

The implications are huge, and many changes seem to be necessary in teachers' education in order to appropriately face the challenge of satisfying all the above-mentioned requests.

Those who have been working in the area and/or who had carefully read Chap. 11 may refute us by saying something like: 'OK, it seems to be a more thoughtful enterprise than we had imagined before, but, at the end of the day, it is just a sophisticated way to say that teachers have to develop their PCK on (or for) modelling.' This kind of refutation opens the way to discussing our second justification for the initial claim made in this section.

From the literature on PCK, independently of the kind of relationship established between subject matter content knowledge and pedagogical knowledge (that is, whether there is an integrative, transformative, or dual-transformative relationship – as previously discussed in Chap. 11), there is no doubt that PCK is the knowledge that teachers mobilise in order to teach particular content. In other worlds, PCK is embedded in specific subject matter. However, *modelling is not a scientific content or topic. Modelling is a way to generate knowledge in science through one's engagement in many epistemic practices*. Therefore, MBT is not a subject matter oriented teaching approach – though it may also support students learning of scientific or socio-scientific topics. As discussed in Chap. 10, the core aims of a learning progression in models and modelling are the development of metacognitive knowledge on models and modelling, and the attainment of competence in modelling – which requires, and implies in, the development of competences in visualisation, drawing and using analogies, argumentation, and understanding about science. The nature of these aims characterises MBT as a being based on the teaching of higher-order skills. And, as in the case of the teaching of other higher-order skills, to identify teachers' needed knowledge as PCK seems inappropriate (Ibraim & Justi, 2015; Zohar, 2004; Zohar & Schwartzer, 2005).

On the other hand, due to the dynamicity and idiosyncrasy of the nature of modelling as being practiced in different domains of experience, for different purposes, or, in the educational context, being taught associated with specific topics at specific school level, it also seems impossible to identify teachers' needed knowledge as "general pedagogical knowledge (that tends to be independent of specific subject matters)" (Zohar, 2004, p. 98). By doing so, the interrelated and knotty network of elements and epistemic practices that characterises modelling as being something so special would be lost. Thus, under the inspiration of Zohar (2004), who originally created a distinct construct for naming teachers' knowledge in the context of the teaching of higher-order thinking, we propose to refer to teachers' pedagogical knowledge for planning and conducting MBT by the phrase *knowledge for teachersbb actions in MBT*. This phrase has two special values: it identifies the main nature of such knowledge – of being one that supports teachers' actions in a special complex context; and it "does not imply a commitment to treat this knowledge as either content-specific or general" (Zohar, 2004, p. 98).

The creation of this phrase does not only answer the core semantic question of 'what is teachers' knowledge of MBT?'. More importantly, it re-dimensions the challenge of educating science teachers as requiring them to become themselves *competent* in the practice of MBT. By doing so, in order to be consistent with its underlying meaning, science teachers' education in this area will have to coherently and simultaneously address a wide variety of issues. This is a challenge that will certainly require a lot of creativity and effort from science educators. Assuming MBT as one of the major ways to support an education for scientific literacy (Chap. 1) and an authentic science education (Chap. 3), we trust that the practitioners in all the sectors identified earlier in this chapter will accept the challenges posed. To retain a claim of being a major sector of education, MBT must succeed.

Concluding Remarks

Finally, we wish to comment on how all these challenges can be met. We have argued that different themes will require different – often subtly different – professional expertise. Whilst coordinating the work of these groups, or at least ensuring communication and collaboration between them, is very difficult, the nearly universal model of all aspects of curriculum development being carried out by one agency is proving dysfunctional. The overall standard of science education across the world is not rising. And it needed to.

In this book, we propose that MBT may be a relevant approach to face these challenges, and we discuss a broad rationale for such a proposal. Our advocacy is for collaborative work focused on an effective inclusion of MBT in science education all distinct school levels (a huge enterprise!). This emerges (i) from our desire to foster conditions for providing twenty-first century citizens with a relevant education in science, and (ii) from our view that the personal development that may follow the processes to be experienced by all agencies and both the individual and collective outcomes reached worth the effort to join such an enterprise.

As we said at the beginning of this book: In general, education has three broad aims. First, it is concerned with the transmission of socially valued knowledge across the generations such that the knowledge acquired by earlier generations is not lost. Second, it seeks to pass on the thinking skills that have produced that knowledge. Third, it supports the production of new knowledge through the use of these skills. The thinking skills involved in the conduct of science in particular are manifest in the processes that lead to scientific knowledge. Models and modelling, therefore, must play important roles in science education if the latter is to be 'authentic', that is to reflect how science has been and should be conducted (Gilbert, 2004).

We look forward to the universal implementation of MBT as a way of simultaneously addressing these aims.

References

Gilbert, J. K. (2004). Models and modelling: Routes to a more authentic science education. *International Journal of Science and Mathematics Education, 2*, 115–130.

Ibraim, S. S., & Justi, R. (2015). *Is PCK a useful construct when pre-service teachers develop their knowledge on argumentation*? Paper presented at the 11th conference of the European Science Education Research Association, Helsinki, Finland.

Zohar, A. (2004). *Higher order thinking in science classrooms: Students' learning and teachers' professional development*. Dordrecht, The Netherlands: Kluwer.

Zohar, A., & Schwartzer, N. (2005). Assessing teachers' pedagogical knowledge in the context of teaching higher-order thinking. *International Journal of Science Education, 27*(13), 1595–1620.

Index

J.K. Gilbert, R. Justi, *Modelling-based Teaching in Science Education*, Models and Modeling in Science Education 9, DOI 10.1007/978-3-319-29039-3

MIX
Papier aus verantwortungsvollen Quellen
Paper from responsible sources
FSC® C105338

If you have any concerns about our products,
you can contact us on
ProductSafety@springernature.com

In case Publisher is established outside the EU,
the EU authorized representative is:
Springer Nature Customer Service Center GmbH
Europaplatz 3, 69115 Heidelberg, Germany

Printed by Libri Plureos GmbH
in Hamburg, Germany